REVERSE PHARMACOLOGY

Phytocannabinoids, Banned and
Restricted Herbals

REVERSE PHARMACOLOGY
Phytocannabinoids, Banned and Restricted Herbals

Amritpal Singh Saroya

Herbal Consultant
Punjab
India

CRC Press
Taylor & Francis Group
Boca Raton London New York

CRC Press is an imprint of the
Taylor & Francis Group, an **informa** business
A SCIENCE PUBLISHERS BOOK

Cover illustrations provided by the author of the book, Amritpal Singh Saroya.

CRC Press
Taylor & Francis Group
6000 Broken Sound Parkway NW, Suite 300
Boca Raton, FL 33487-2742

First issued in paperback 2020

© 2018 by Taylor & Francis Group, LLC
CRC Press is an imprint of Taylor & Francis Group, an Informa business

No claim to original U.S. Government works

ISBN-13: 978-1-138-03707-6 (hbk)
ISBN-13: 978-0-367-78131-6 (pbk)

Library of Congress Cataloging-in-Publication Data

Names: Amritpal Singh, 1971- author.
Title: Reverse pharmacology : phytocannabinoids, banned and restricted herbals / Amritpal Singh Saroya.
Description: Boca Raton, FL : CRC Press, 2017. | "A Science Publishers book."
| Includes bibliographical references and index.
Identifiers: LCCN 2017021748| ISBN 9781138037076 (hardback : alk. paper) |
ISBN 9781315178059 (e-book)
Subjects: | MESH: Plant Preparations--pharmacology |
Nanoparticles--therapeutic use | Drug Discovery--methods |
Pharmacognosy--methods | Cannabinoids--pharmacology
Classification: LCC RS164 | NLM QV 766 | DDC 615.3/21--dc23
LC record available at https://lccn.loc.gov/2017021748

Visit the Taylor & Francis Web site at
http://www.taylorandfrancis.com

and the CRC Press Web site at
http://www.crcpress.com

Preface

The book titled REVERSE PHARMACOLOGY Phytocannabinoids, Banned and Restricted Herbals has been prepared on the persistent demand of the herbal drug industry. Part A deals with *REVERSE PHARMACOLOGY & NANOPHYTOMEDICINE*. The linkage of Reverse Pharmacology with Phytomedicine and Ayurvedic medicine has been discussed. Nanocurcumin, nanoandrographolide, nanohypericum, nanohypericin and nanohyperforin, nanotaxol, nanosilymarin and nanoparticles of other bioactive compounds and herbal extracts have been discussed to examine pharmacological aspects and possible application in therapeutics.

PART B: *PHYTOCANNABINOIDS* deals with botany, chemistry, pharmacology and toxicology of *Cannabis indica, C. sativa* and *C. ruderalis*. A chapter on Herbal Cannabinomimetics is the salient feature of this section. Separate chapters have been devoted to Cannabis oil and *Leonotis leonurus* (Wild Cannabis) explaining the chemistry and pharmacological investigations. A–Z of the restricted and banned herbal drugs is integral part of the part b. This section is of particular importance for regulatory affairs of restricted and banned herbs of commerce.

I sincerely hope the book shall be welcomed by the herbal fraternity and serve as a handy collection in ever expanding market of herbal drugs. Suggestions and critics for improving the quality of the book are welcome.

Contents

PART A

REVERSE PHARMACOLOGY & NANOPHYTOMEDICINE

CHAPTER 1

Introduction

1.1 What is Reverse Pharmacology?

Reverse and forward pharmacology are two paths generally adopted for the purpose of the drug-discovery (Takenaka 2001). The difference between reverse and forward pharmacology is depicted in the following diagram (Fig. 1).

Reverse pharmacology (RP) has also been addressed as target-based drug discovery (Lee et al. 2012). As per standard definition available, reverse pharmacology is defined as the science integrating documented clinical/experiential hits, into leads by transdisciplinary exploratory studies and further developing these into drug candidates by experimental and clinical research. The reverse pharmacognosy is aimed at finding novel biological targets for discovering natural compounds (Ernst 2009). This may involve virtual or real screening and identification of the natural resources that may be source of the active molecules (Do et al. 2005).

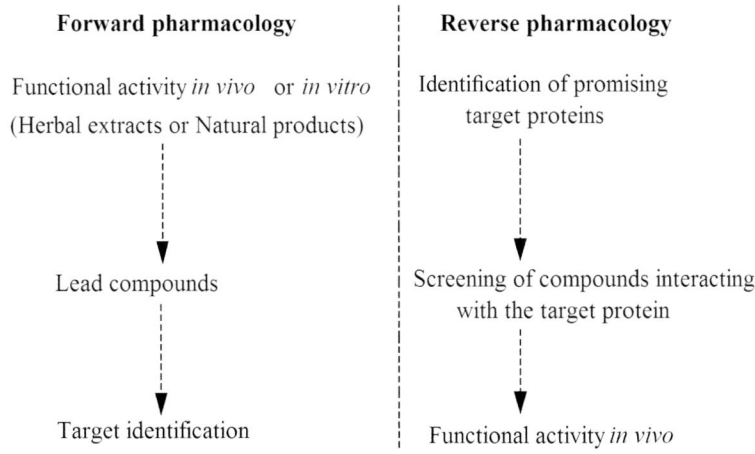

Forward pharmacology

Functional activity *in vivo* or *in vitro*
(Herbal extracts or Natural products)

↓

Lead compounds

↓

Target identification

Reverse pharmacology

Identification of promising
target proteins

↓

Screening of compounds interacting
with the target protein

↓

Functional activity *in vivo*

Fig. 1. Forward and reverse pharmacology approaches in drug-discovery.

1.2 Reverse Pharmacognosy

Forward pharmacology is also called classical pharmacology. Pharmacognosy, the science dealing with study of physical, chemical and biological properties, cultivation and storage of herbal drugs is essentially viewed as classical pharmacology. The sciences of pharmacognosy and classical pharmacology are generally compared with reverse pharmacology (see definition).

In pharmacognosy, the folk medicines undergo clinical testing for possible efficacy. Once the efficacy is established, studies are done so as to study the biological target of the drug. This approach or path defines the classical pharmacognosy.

In terms of a routine protocol, preliminary phytochemical screening is done and the characterisation and isolation of the active ingredient (active constituent or bioactive) is done. When we talk about the routine protocol of reverse pharmacognosy, knowledge of a key enzyme being affected by a bioactive is vital. This approach or path can identify medicinal plants as source of novel bioactives (Do and Bernard 2004).

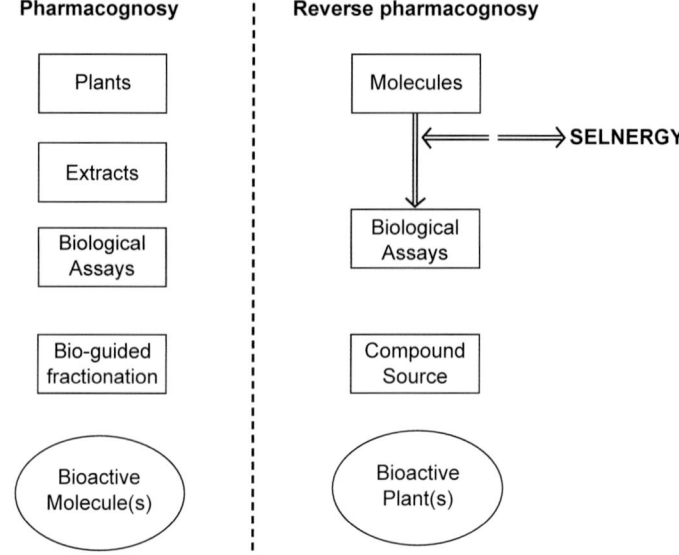

Fig. 2. Pharmacognosy and reverse pharmacognosy approaches in drug-discovery.

Reverse pharmacognosy is a potential tool for addressing certain issues in current drug discovery process. A few to mention are scarcity of hits for clinical purpose and toxicology data by exploitation of existing pharmacognosy data (Do et al. 2005; Blondeau et al. 2016). Integration of pharmacognosy and reverse pharmacognosy for the purpose of research can be a handy and fast-flowing tool for drug discovery from natural resources (Ernst 2009).

1.3 Viniferin

Epsilon-viniferin or ε-viniferin (Fig. 3), is employed as an active ingredient in cosmetic products. This compound is basically a phenolic compound and occurs in *Vitis coignetiae* Pulliat ex Planch. (crimson glory vine), *Vitis vinifera* Linn. (grapevine) and *Dryobalanops aromatica* Gaertn (camphor tree).

Despite wide usage in cosmetology, biological properties of viniferin are poorly understood. Selnergy™ is a novel tool employed for the purpose of drug discovery. It consists of a docking software which predicts interaction of ligands with proteins. Selnergy has been used for identification of binding targets for viniferin.

As far as cosmetic application is concerned, cyclic nucleotide phosphodiesterase 4 (PDE4) was the most promising candidate. The others PDE1, 2, 3, 5 and 6 subtypes were not retained, which clearly indicates for a selectivity for PDE4. Investigatory tests with 6 subtypes of phosphodiesterase demonstrated a significant selectivity for phosphodiesterase 4 subtype. Evaluation of ε-viniferin on the secretion of markers of inflammation confirmed the selectivity (Do et al. 2005).

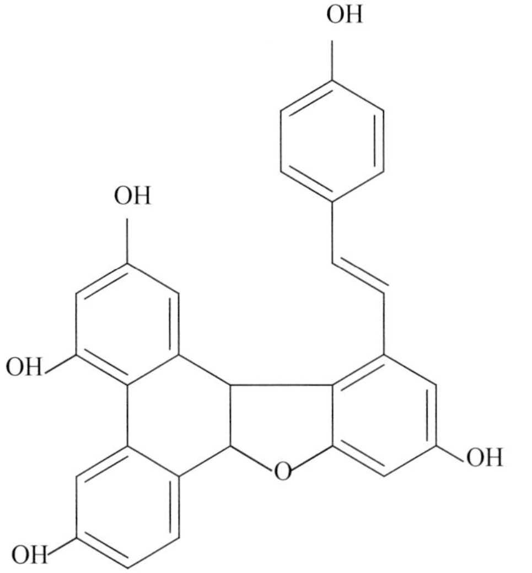

Fig. 3. Chemical structure of viniferin.

1.4 Meranzin

Meranzin (Fig. 4) is a coumarin compound isolated from *Limnocitrus littoralis* (Miq.) Swingle (Limnocitrus). This plant is endemic to Central Java. Like ε-viniferin, Selnergy™ was used for identification of putative binding targets of the coumarin. Cyclooxygenase 1 (COX1), Cyclooxygenase 2 (COX2) and peroxisome proliferator-activated receptor (PPAR gamma) were the targets selected. The results obtained from the software were comparable with the experimental findings. This proved that an extract of the herbal drug containing meranzin modulated COX1, COX2 and PPAR gamma (Do et al. 2007).

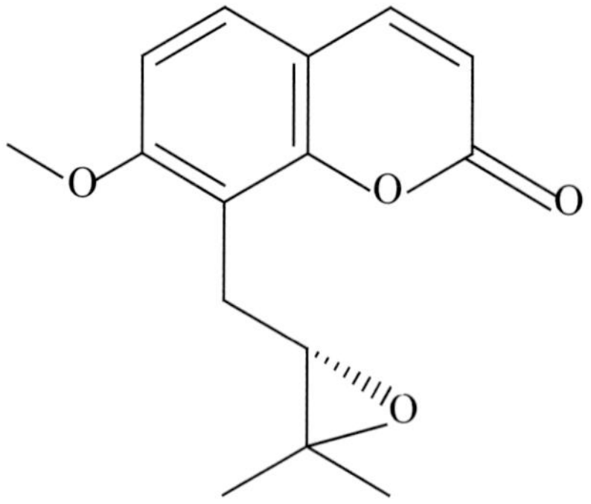

Fig. 4. Chemical structure of meranzin.

Further Reading

Blondeau S, Do QT, Scior T, Bernard P, Morin-Allory L. Reverse pharmacognosy: another way to harness the generosity of nature. *Curr Pharm Des*. 2010; **16:** 1682–96.

Do QT, Bernard P. Pharmacognosy and reverse pharmacognosy: a new concept for accelerating natural drug discovery. *Drugs*. 2004; **7:** 1017–27.

Do QT, Renimel I, Andre P, Lugnier C, Muller CD, Bernard P. Reverse pharmacognosy: application of selnergy, a new tool for lead discovery. The example of epsilon-viniferin. *Curr Drug Discov Technol*. 2005; **2:** 161–7.

Do QT, Lamy C, Renimel I, Sauvan N, André P, Himbert F, Morin-Allory L, Bernard P. Reverse pharmacognosy: identifying biological properties for plants by means of their molecule constituents: application to meranzin. *Planta Med*. 2007; **73:** 1235–40.

Ernst E. Reverse pharmacognosy: Editorial. *J Diet Suppl*. 2009; **6**(3): 201–202.

Lee JA, Uhlik MT, Moxham CM, Tomandl D, Sall DJ. Modern phenotypic drug discovery is a viable, neoclassic pharma strategy. *J Med Chem*. 2012; **55:** 4527–38.

Takenaka T. Classical vs reverse pharmacology in drug discovery. *BJU Int*. 2001; **88:** 49–50.

Vaidya BD. Reverse pharmacology—A paradigm shift for drug discovery and development. *Curr Res Drug Discov*. 2014; **1:** 39–44.

CHAPTER 2

Reverse Pharmacology and Phytomedicine

2.1 The Decoction of *Argemone mexicana* for the Treatment of Uncomplicated Falciparum Malaria

An exceptional, dose-escalating clinical trial conducted in Mali investigated the efficacy of a decoction of *Argemone mexicana* L. (Papaveraceae) in the treatment of uncomplicated malaria caused by *Plasmodium falciparum*. A dose-escalating clinical trial refers to a progressive increase in the amount of any drug. The purpose is either to improve the tolerability or to maximize the effect.

The eighty patients enrolled for the clinical trial. Majority of the patients (75%) were in the age group of less than five years. Rest of the patients (25%) were less than one year. The patients were assigned to consume the dose of decoction of *A. mexicana* in three regimens as follows:

Dose patterns of a decoction of *A. mexicana* in a dose-escalating clinical trial.

S. No.	Group N	Duration	Dose
1.	A 23	3 days	OD
2.	B 40	7 days	BD
3.	C 17	7 days	QD (4 days) followed by BD (3 days)

The eighty patients diagnosed with *P. falciparum* malaria has parasitaemia > 2000/ microl. As far as clinical response was concerned, it was 35%, 73% and 65% in three groups, respectively. On the 14th day, whole dimensions of the adult:child ratio were found to be lower in case of children aged < one year. It was higher in case of patients aged > 5 years. A complete clearance of *P. falciparum* was not achieved. However, 67% of the patients achieved parasitaemia < 2000/microl (Willcox et al. 2007).

2.2 *A. mexicana* Decoction versus Artesunate-amodiaquine for the Management of Malaria in Mali

A study evaluated the efficacy of a decoction of *Argemone mexicana* in 301 patients diagnosed with malaria in age group of five years. The patients were randomised to receive the herbal decoction or standard antimalarials (artesunate-amodiaquine). The

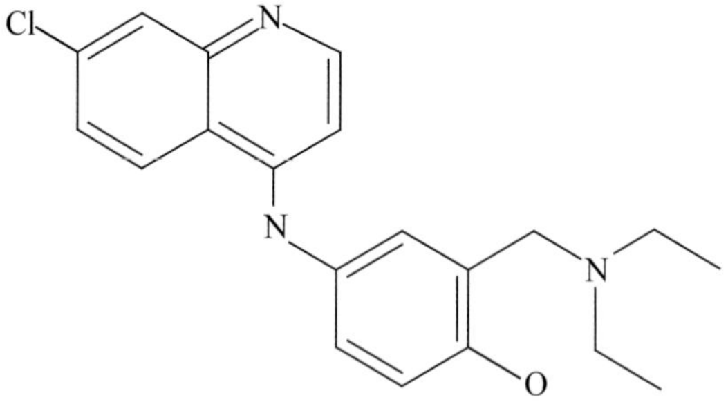

Fig. 5. Chemical structure of amodiaquine.

patients tolerated both the drugs. After a treatment of twenty-eight days, alternative treatment was not required in 89% of the patients receiving the herbal decoction. In contrast, 95% of the patients receiving artesunate-amodiaquine responded well (Graz et al. 2010).

2.3 Reverse Pharmacology Approach Applied to the Decoction of *A. mexicana*

Three alkaloids, allocryptopine, berberine and protopine were isolated from *A. mexicana* through bioguided fractionation. Allocryptopine (Fig. 6), protopine (Fig. 7) and berberine (Fig. 8) demonstrated antimalarial activity *in vitro*. Allocryptopine and protopine showed selective response towards the parasite, *Plasmodium falciparum*.

The study emphasised the need of pharmacokinetic studies in order to determine whether the three alkaloids in question can be utilised as marker compounds for the purpose of quality control and standardisation of the decoction of *A. mexicana* (Simoes-Pires et al. 2014).

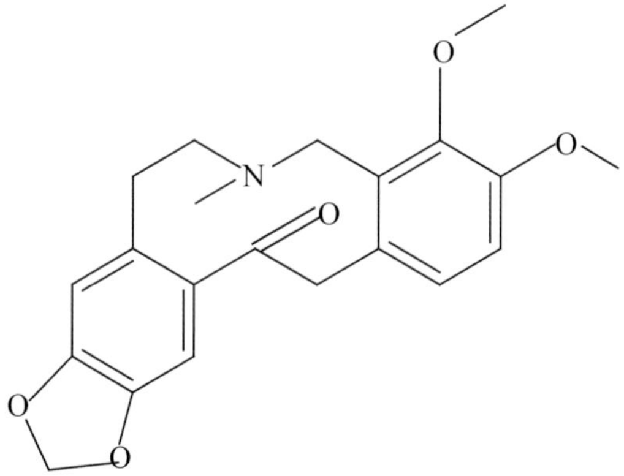

Fig. 6. Chemical structure of allocryptopine.

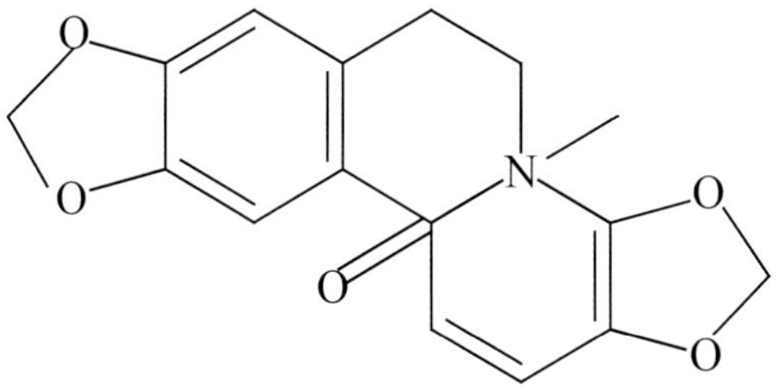

Fig. 7. Chemical structure of protopine.

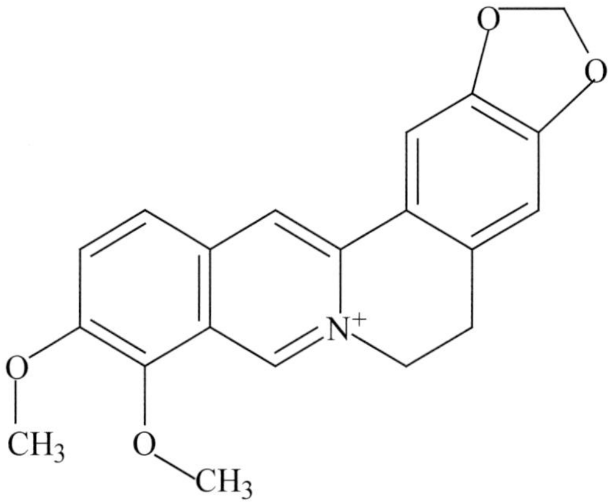

Fig. 8. Chemical structure of berberine.

Further Reading

Graz B, Willcox ML, Diakite C, Falquet J, Dackuo F, Sidibe O, Giani S, Diallo D. *Argemone mexicana* decoction versus artesunate-amodiaquine for the management of malaria in Mali: policy and public-health implications. *Trans R Soc Trop Med Hyg.* 2010; **104**(1): 33–41.

Simoes-Pires C, Hostettmann K, Haouala A, Cuendet M, Falquet J, Graz B, Christen P. Reverse pharmacology for developing an anti-malarial phytomedicine. The example of *Argemone mexicana*. *Int J Parasitol Drugs Drug Resist.* 2014; **4**(3): 338–46.

Willcox ML, Graz B, Falquet J, Sidibé O, Forster M, Diallo D. *Argemone mexicana* decoction for the treatment of uncomplicated falciparum malaria. *Trans R Soc Trop Med Hyg.* 2007; **101**(12): 1190–8.

CHAPTER 3

Reverse Pharmacology and Ayurvedic Medicine

3.1 Definition

Reverse pharmacology is the science of integrating documented clinical/experiential hits, into leads by transdisciplinary exploratory studies and further developing these into drug candidates by experimental and clinical research (Figs. 9 and 10).

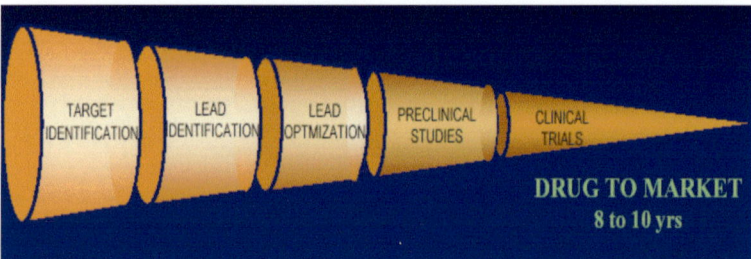

Fig. 9. Drug development with conventional pharmacology.

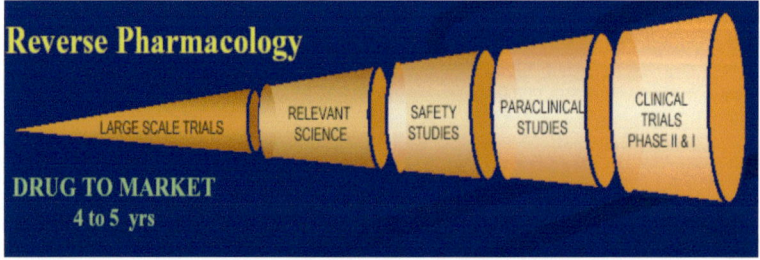

Fig. 10. Drug development with reverse pharmacology.

3.2 Scope of Reverse Pharmacology

The scope of reverse pharmacology is to understand the mechanisms of action at multiple levels of biological organisation and to optimise safety, efficacy and acceptability of the leads in natural products, based on relevant science (Vaidya 2006).

3.3 Dimensions of Reverse Pharmacology

- Experiential documentation
  - Pharmacoepidemiology—Standardised formulation with HPLC pattern
- Exploratory human/animal studies—Relevant models of activity—Human dose determination
- Experimental programmes
  - Levels of biological organisation
  - Rapid drug development path
  - Leads: Comb. Leads: Combinatorial Chemistry Chem & HTPS.

3.4 Examples

- *Mucuna pruriens* for Parkinson's disease
- *Zingiber officinale* for nausea/vomiting
- *Picrorhiza kurroa* for Hepatitis
- *Curcuma longa* for Oral Cancer
- *Panchavalkal* for burns and wounds
- *Azadirachta indica* for Malaria

3.5 *Mucuna pruriens* for Parkinson's Disease

L-dopa content: 2.5 g%–12.0 g% (Damodaran and Ramaswamy 1937). Other phytochemicals: mucunain, pruriendine, 5-HTP, aminoacids, dopaquinones, melanin, alanine, linolenic, methionine, and niacin, etc. *M. pruriens* seeds and HP-200 standardised for L-dopa content (3.9–4.2% and 3.0–3.7%, respectively). A study by Vaidya et al. (1978) reported the role of *M. pruriens* in the treatment of Parkinson's disease (Fig. 11).

3.6 *Zingiber officinale* for Nausea/Vomiting

Modern-day uses of ginger in Eastern medicine include the use of the herb to treat nausea (including motion sickness and morning sickness during pregnancy). The main components of ginger are the aromatic essential oils, anti-oxidants, and the pungent oleo-resin. These aromatic or pungent compounds have been identified as phenylalkylketones, known as gingerols, shogaols, and zingerone.

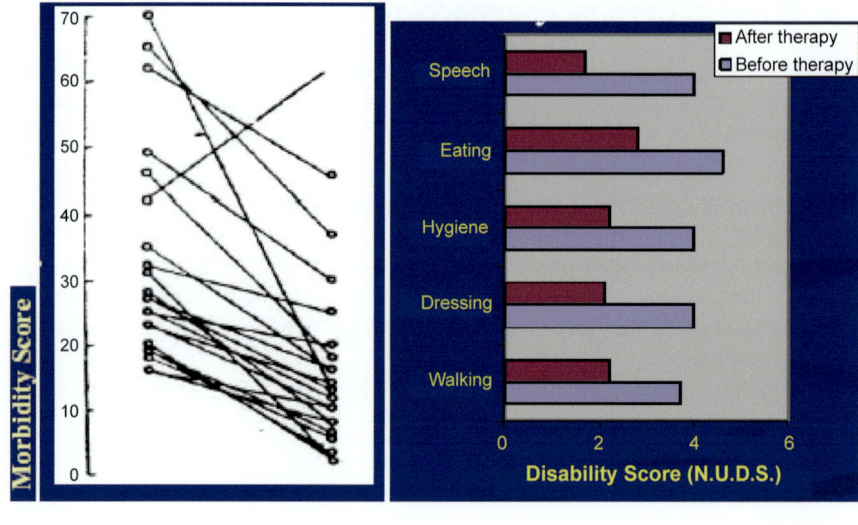

38.30±3.55   17.84±3.12

Mean ±S.E.

Fig. 11. Clinical activity of *M. pruriens* in Parkinson's disease.

Zingerone (Fig. 12), diterpenoid constituent of ginger has been shown to have activity similar to a 5HT3 antagonist, ondesteron and other anti-emetic drugs used as adjuncts to chemotherapy. 5HT3 receptors are found in both the chemoreceptor trigger zone and on the vagal nerve terminals in the intestine. The anti-emetic effects of ginger are due to its local effect on the vagal receptors in the stomach.

A double-blind randomised placebo-controlled trial was carried to investigate the effect of ginger extract on symptoms of morning sickness. The participants included 120 women who were less than 20 weeks pregnant and had experienced morning sickness daily for at least a week. They have had no relief of symptoms through

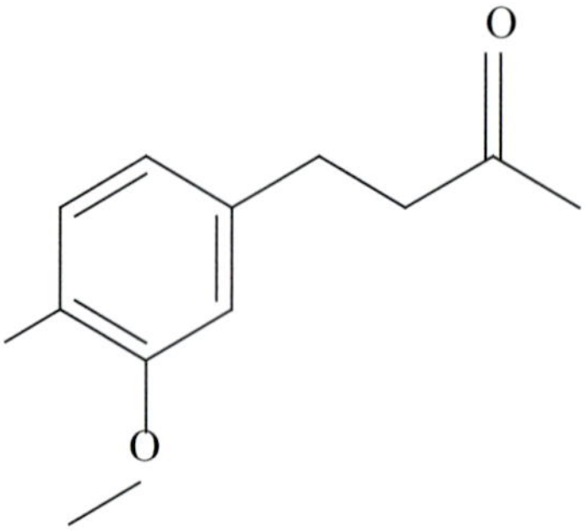

Fig. 12. Chemical structure of zingerone.

dietary changes. Random allocation of 125 mg ginger extract equivalent to 1.5 g of dried ginger was made or placebo was given four times per day for 4 days. The nausea experience score was significantly less for the ginger extract group in relation to the placebo group after the first day of treatment. This difference continued to be present on each treatment day (Arfeen et al. 1995).

3.7 *Azadirachta indica* for Malaria

There are number of references in classical Ayurvedic texts for neem as antimalarial agent. Classical formulations for Vishamajwara include Nimbadi Kwath, Amruta Nimba Kwath, Panchatiktaka Ghrita, and Jwarhara Kwath. Neem is still widely used in India. *In vitro* studies with limonoids isolated from neem extract have shown efficacy against Plasmodium species (Table 1). Nimbolide (Fig. 13) was identified as an antiplasmodial compound (IC50 = 0.95 ng ml^{-1}, *P. falciparum* K1). The derivatives nimbinin and gedunin (Fig. 13) and its dihydro-derivative was also found to be active *in vitro* against Plasmodium parasites with EC50 values from 0.72 to 1.74 lg ml^{-1} (MacKinnon and Read 1999).

Table 1. Efficacy of limonoids of neem in Plasmodium species.

Limonoid	IC50 (μmol/l)	Falciparum Clone	Reference
Gedunin	0.200	W2(CQ-9)	MacKinnon and Read 1999
Gedunin	0.039	D6(CQ-S)	MacKinnon and Read 1999
Nimbinin	0.77	K1(MDR)	Bray et al. 1990
Nimbolide	2.0	K1(MDR)	Rochanakij et al. 1985
Nimbin	50	K1(MDR)	Bray et al. 1990

3.8 Reverse Pharmacology and Ayurveda

Reverse pharmacology correlates with Ayurvedic drug action. The approach has attracted sizeable attention, nationally and internationally. Central Scientific Instrumental Research (CSIR) and Indian Council of Medical Research (ICMR) have done clinical trials with natural products. Moreover, Central Council for Research in Ayurveda and Siddha (CCRAS) has recently adopted the golden triangle approach for some new indications of old drugs, as well as for Ayurveda (Fig. 14).

Golden triangle approach is a combination of Dravyagunavignyan, system biology, and reverse pharmacology for discovering potent and cost-effective remedies. Dravyaguna deals with the Ayurvedic study of drugs derived from natural (plant, animal or marine) origin. Dravya refers to a constituent of the universe and guna signifies property.

In the recent past, the study of Dravyaguna has become more important because of the global acceptance of Ayurvedic system of medicine. Ayurveda has its own concept as far as drug formulation is concerned. Ayurveda has documented Dravya, Guna,

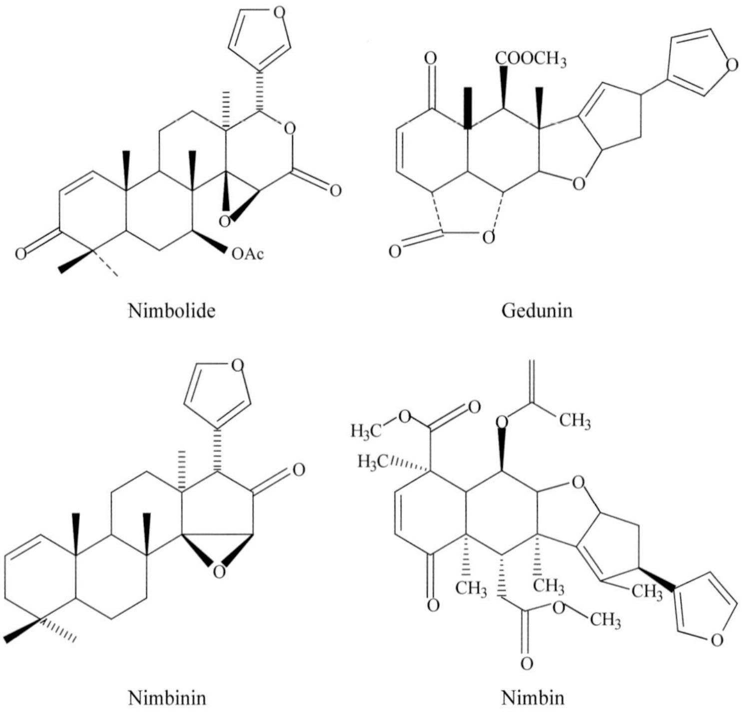

Fig. 13. Antimalarial limonoids of neem.

Rasa, Virya, Vipaka, Prabhava and Karma as medicinal agents and these represent the pharmacological aspects of drug usage in Ayurveda.

In Ayurveda, drugs have been classified in a number of ways but the classification based on the action of the drug is widely accepted. For instance, a drug used for alleviating worm infection is known as anthelmintic and in Ayurvedic language, krimighana.

In modern medicine, drugs have been classified according to their pharmacological actions. Medicinal plants such as Ashwagandha, Brahami, Tulsi, Guggul, Kutki, Kalmegha, Gokshura and Shatavari have been identified by the modern medical science for application in medicine. Active constituents of the plants have been identified and highly purified extracts are being marketed.

Studying the Ayurvedic drugs at biomolecular levels may unravel the mechanism of action which has eluded scientists for a long time. Hepatology and rheumatology are the two areas where Ayurvedic remedies are even prescribed by Allopathic Physicians.

Silybum marianum is a well-documented western herbal remedy that is used for treating liver diseases. In India, we have *Picrorhiza kurroa* (kutki), which is a priced drug for treating liver diseases. *P. kurroa*, when compared with *S. marianum*, the hepatoprotective effect of *P. kurroa* was found to be similar, or in many cases, superior to the effect of *S. marianum*. But *S. marianum* seems to be more popular

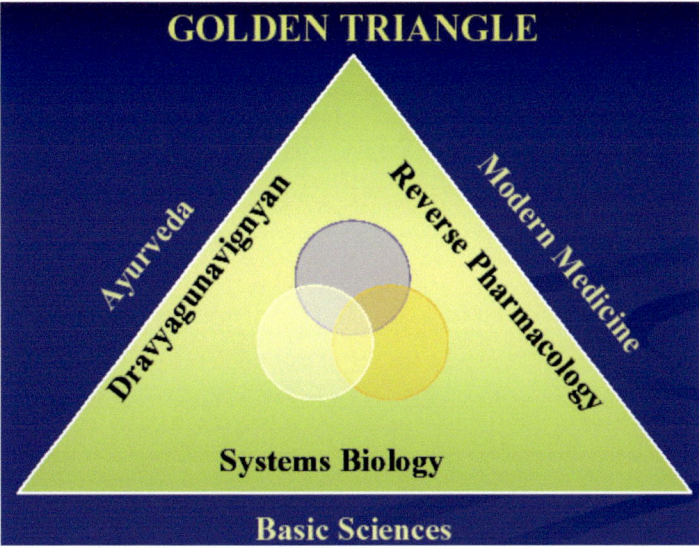

Fig. 14. Golden triangle approach.

than *P. kurroa*. Silymarin, the active constituent of *S. marianum,* has been isolated and purified. Moreover, complete pharmacological and pharmacokinetic data is available for the drug.

The reverse approach in pharmacology has been quite successfully applied in the past. The major drawback was the long time lag from the observational therapeutics to a new drug. For example, *Rauwolfia serpentina* (sarpagandha) was convincingly demonstrated to be anti-hypertensive by Sen and Bose in 1931. But a drug reserprine, emerged only after 20 years of work by Vakil 1949. This happened because the path of reverse pharmacology was random and quite discontinous.

The paradigm of reverse pharmacology is actually a rediscovery of the path, which founded modern pharmacology. Table 2 lists the names of plants, clinical effects, and experimental correlates that illustrate how novel clinical bio-dynamic effects can lead to the development of basic disciplines in pharmacology and biology.

Table 2. Re-discovery of the paradigm of reverse pharmacology.

Medicinal Plant	Clinical Effect	Experimental Correlate
Chondrodendron tomentosum	Paralysis and death	Neuromuscular block
Cinchona officinalis	Fever	Antimalarial
Digitalis purpurea	Dropsy	Na+-K+ ATPase
Papaver somniferum	Analgesia	Opioid receptors
Physostigma venenosum	Ordeal poison	Anticholinesterase
Salix alba	Fever and pain	Prostaglandins
Strychnos nux-vomica	Stimulant and convulsant	Glycinergic receptors

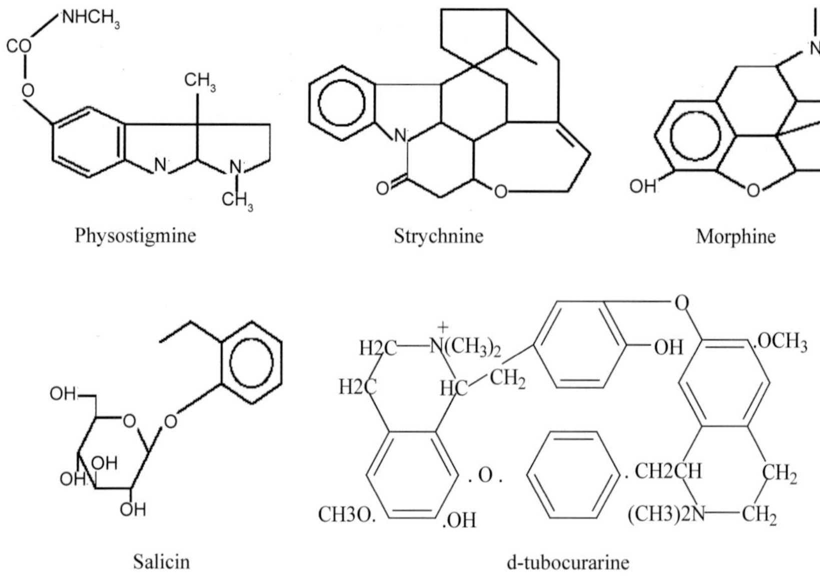

Fig. 15. Active constituents of plants listed in Table 2.

There has been a renaissance in Ayurvedic research that the western and Indian pharma companies have just begun to notice. Reverse pharmacology was only sporadically applied to new drug development. It is the need of the hour to document the unknown, the unintended and the desirable novel prophylactic and therapeutic effects in observational therapeutics. Several new classes of drugs have accidently emerged due to this path (Fig. 16).

3.9 Discovery of Reserpine

Reserpine, an alkaloid is purified from the roots of *Rauwolfia serpentina*. Reserpine is credited as the first potent drug ever used in the long-term treatment of high blood pressure (hypertension). Extracts of *R. serpentina* (popularly known as Sarpagandha) have been used in Ayurveda for treating snakebite, insomnia and insanity.

Rumpf first reported about rauwolfia in his Herbarium amboinense in the year 1755. Sen and Bose described the therapeutic utility of *R. serpentina* in the Indian Medical Journal in 1931. Vakil enunciated the role of *R. serpentina* in the treatment of hypertension in the British Heart Journal for the first time (Goenka 2007).

Müller et al. (1952) reported the detailed chemistry and pharmacology of *R. serpentina*. Serpasil, the commercial preparation of reserpine was introduced in 1952 for the treatment of hypertension, tachycardia and thyreotoxicosis (Lobay 2015). In psychiatry, reserpine was prescribed as a tranquilising agent until modern synthetic antidepressant and antipsychotic drugs (Fig. 18) were introduced (López-Muñoz et al. 2004; Jerie 2007).

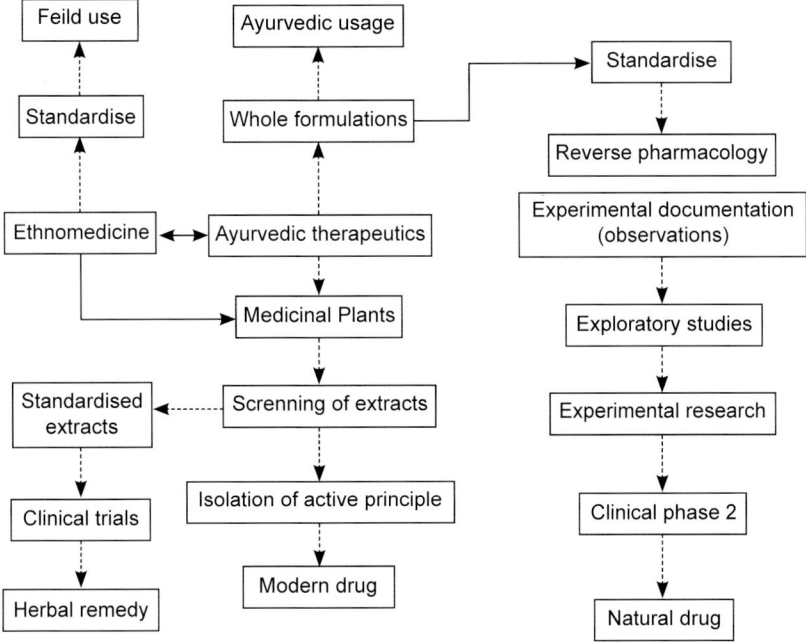

Fig. 16. Scheme for drug discovery from plants used in Ayurveda.

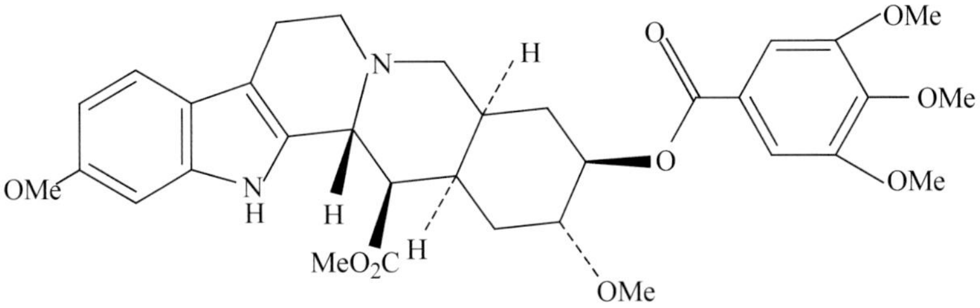

Fig. 17. Structure of reserpine.

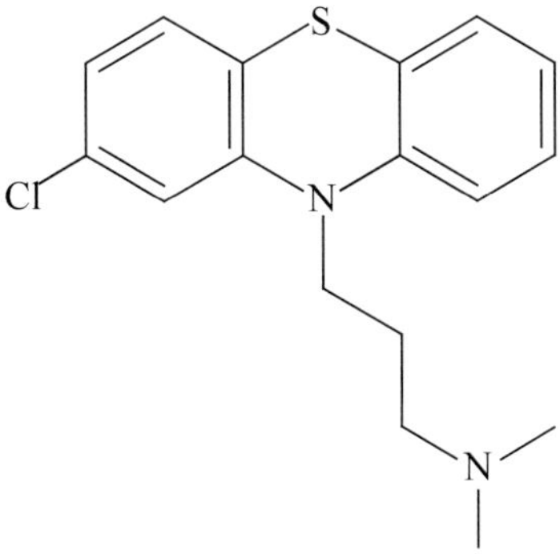

Fig. 18. Structure of chlorpromazine.

Further Reading

Arfeen Z, Owen H, Plummer JL, Ilsley AH, Sorby-Adams RA, Doecke, CJ. A double-blind randomised controlled trial of ginger for the prevention of postoperative nausea and vomiting. *Anaesthsia Intensive Care.* 1995; **23:** 449–452.

Bray DH, Connolly JD, Peters W, Phillipson JD, Robinson BL, Tella A, ThebtarAnonth Y, Warhurst DC, Yuthavong Y. Antimalarial activity of some Limonoids. *Transactions Royal Soc Trop Med Hyg.* 1985; **73:** 426.

Damodaran M, Ramaswamy R. Isolation of L-Dopa from the seeds of *Mucuna pruriens. Biochemistry.* 1937; **31:** 2149.

Goenka AH. Rustom Jal Vakil and the saga of *Rauwolfia serpentina. J Med Biogr.* 2007; **15:** 195–200.

Jerie P. Milestones of cardiovascular therapy. IV. Reserpine. *Cas Lek Cesk.* 2007; **146:** 573–7.

Lobay D. Rauwolfia in the Treatment of Hypertension. *Integr Med (Encinitas).* 2015; **14:** 40–6.

López-Muñoz F, Bhatara VS, Alamo C, Cuenca E. Historical approach to reserpine discovery and its introduction in psychiatry. *Actas Esp Psiquiatr.* 2004; **32:** 387–95.

MacKinnon MJ, Read AF. Genetic relationships between parasite virulence and transmission in the rodent malaria *Plasmodium chabaudi. Evolution.* 1999; **53:** 689–703.

MacKinnon S, Durst T, Arnason JT, Angerhofer C, Pezzuto J, Sanchez-Vindas PE, Poveda LJ, Gbeassor M. Antimalarial activity of tropical Meliaceae extracts and gedunin derivatives. *J Nat Prod.* 1997; **60:** 336–41.

Müller JM, Schlittler E, Bein, HJ. Reserpin, der sedative Wirkstoff aus *Rauwolfia serpentina* Benth. *Experientia.* 1952; **8:** 338.

Rochanakij S, Thebtaranonth Y, Yenjai C, Yuthavong Y. Nimbolide, a constituent of Azadirachta indica, inhibits *Plasmodium falciparum* in culture. *Southeast Asian J Trop Med Public Health.* 1985; **16:** 66–72.

Sen G, Bose K. Rauwolfia serpentina & new Indian drug industry and high blood pressure. *Indian Med Wld.* 1931; **2:** 194.

Singh AP. *Dravyaguna Vijnana.* Chaukhambha Orientalia, New Delhi, 2005.

Vaidya AB et al. Treatment of Parkinson's diseases with cowhage plant–*Mucuna pruriens* Bak. *Neurol India.* 1978; **4:** 171–176.

Vaidya ADB. Reverse pharmacological correlates of Ayurvedic drug action. *Indian Journal of Pharmacology.* 2006; **38:** 311–315.

Vaidya ADB, Raut AK, Amonkar AJ. *Active Principle Identification and Determination.* USP-IP 6th Ann. Meeting, 2007.

Vakil RJ. A clinical trial of Rauwolfia serpentina in essential hypertension. *Brit Heart J.* 1949; **11:** 350–355.

CHAPTER 4

Introduction to Nanophytomedicine

4.1 Introduction

In the recent past, significant advancement has been made on developing novel drug delivery systems (NDDS) for phytodrugs (herbal drugs) and extracts. Animations, ethosomes, liposomes, microsphere, nanocapsules, phytosomes, polymeric nanoparticles, and transferosomes are some of the novel formulations that have evolved using herbal bioactives and herbal extracts. These novel NDDS have been reported to have significant advantages over conventional drug delivery systems in herbal medicine including expressed juice, powder, decoction, infusion and tincture.

The points of advantage includes in terms of increased solubility and bioavailability, reduced incidence of toxicity, enhanced pharmacological activity and stability (a critical as well as a limiting factor for herbal drugs), improved kinetics and delivery at target organ. Above all, the novel NDDS offer significant protective effect on degradation of physical and chemical origin (Ajazuddin 2010).

The therapeutic potential of phytodrugs is addressed as phytotherapeutics. This science definitely requires a scientific methodology to ensure sustainable delivery of the essential components (active constituents) of herbal drugs or formulations so as to increase efficacy among patients and to prevent repeated administration.

Phytotherapeutics need a scientific approach to deliver the components in a sustained manner to increase patient compliance and avoid repeated administration. This can be achieved by designing novel drug delivery systems (NDDS) for herbal constituents. NDDS for herbal drugs in addition to above described points, reduce the incidence of toxicity and improves the bioavailability, a factor critical for phytotherapeutics. Nanotechnology is an novel and important tool for achieving this ambitious aim. Nanophytomedicine has a definite future as it can prove handy for overcoming problems associated with traditional NDDS for phytodrugs (Ansari et al. 2012).

The use of nanotechnology for treatment, identification, monitoring, and managing biological systems has recently been named as the field of 'nanomedicine'. In the herbal formulation research, incorporating nano-based formulation has a number of advantages for phytomedicine, including improvement of solubility and bioavailability, safeguard from toxicity, enhancement of pharmacological activity, improvement of stability, and increase in tissue macrophages distribution, sustained delivery, and protection from physical and chemical degradation (Patravale et al. 2015).

It has been widely proposed to combine herbal medicine with nanotechnology, because the nanostructured systems might be able to potentiate the action of plant extracts, reducing the required dose and side effects, and improving activity. Nanosystems can deliver the active constituent at a sufficient concentration during the entire treatment period, directing it to the desired site of action. Conventional treatments do not meet these requirements.

4.2 Nanomedicine and Nanophytomedicine

Nanomedicine is defined as the science involving the application of nanotechnology for the purpose of treatment, identification, monitoring, and managing biological systems. In the field of phytomedicine research, incorporation of the nanotechnology has a significant advantages to offer (Patravale et al. 2015).

It has been postulate that in nanophytomedicine, nanostructured systems might be potentiating the action of plant extracts, reducing the required dose and side effects, and improving activity. Nanophytomedicines have the potential in delivering the active constituent in a adequate concentration during the entire treatment, thereby directing it to the desired site of action. In comparison, the conventional phytomedicines, lack these requirements.

4.3 Advantages of Nanophytomedicine

Nanophytomedicines have several advantages (Patravale et al. 2015), which has been summarized below and demonstrated in Fig. 19.

- Enhanced pharmacological activity,
- Enhanced stability,
- Enhancement of solubility,
- Improved bioavailability,

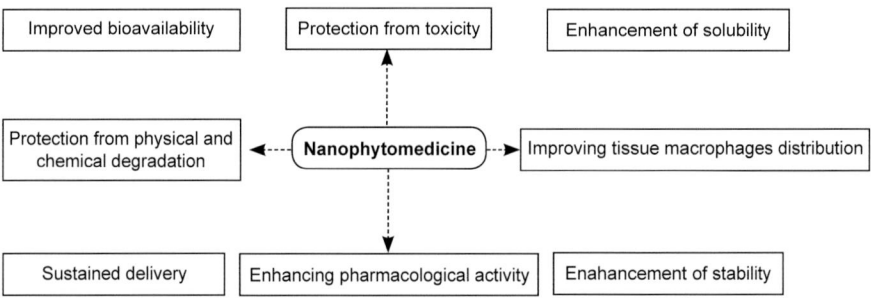

Fig. 19. Diagrammatic representation of advantages of nanophytomedicine (Adapted from Gunasekaran et al. 2014).

- Protection from physical and chemical degradation,
- Protection from toxicity,
- Sustained delivery.

4.4 Nanophytomedicine and Homeopathy

Homeopathy is an alternative system of medicine based on the fundamental principle 'like cures like'. In several incidences, homeopathic clinicians have made claims of successful treatment of life-threatening conditions like cancer. Preclinical studies have been done to verify these claims. On basis of the data gathered from controlled studies, it has been found that Homeopathic medicines have modulating effect on the gene expression and biological signalling pathways that regulate the cell cycles. Homeopathic medicines have also impact on immune and central nervous system as proved in *in vivo* and *in vitro* studies.

By utilizing services of nanotechnology, not only the Homeopathic manufacturing can be improved but characterization of nanoparticle end products can be done. Further, nanotechnology can be handy in describing the interactions of Homeopathic nanomedicine with the living system. This shall have rapid impact on cost-effectiveness and safety of personalized (Bell et al. 2015).

4.5 Nanotechnologies Approaches for Enhancing the Bioavailability and Bioactivity of Phytomedicine

4.5.1 *Reducing the size of the phytomedicine into nanophytomedicine* (Gunasekaran et al. 2014)

The modes of administration of drugs in phytomedicine are limited as compared to conventional medicine. The oral route is the preferred method for the administration of drugs in phytomedicine. Convenience on the part of the patient and superiority as compared to other modes of administration gives the added advantage to the oral route.

As far the oral route of administration of drugs is concerned, a few of rate-determining steps have significant role in the absorption of the drugs. The first rate-determining step is the rate of dissolution. Another rate-determining step is the rate of the permeation of the drug through the membrane.

Dissolution is a critical rate-determining step for hydrophobic drugs. The hydrophobic drugs are poorly soluble in water. On the other hand, the hydrophilic drugs have high solubility in water and thus the dissolution is rapid. In case of hydrophilic drugs, the rate of the permeation through the membrane is the rate-determining step. This is known as transmembrane rate limited.

4.5.2 *Modification of surface properties* (Gunasekaran et al. 2014)

The modification of the surface is useful for enhancement of the permeation through a bio-membrane. When there is passage of phytomedicine through the gut (intestine),

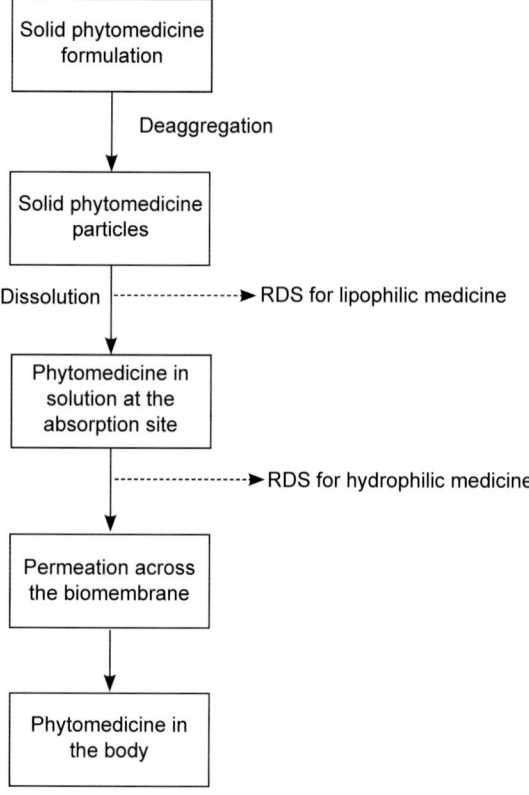

Fig. 20. Reducing the size of the phytomedicine into nanophytomedicine (Adapted from Gunasekaran et al. 2014).

there occurs a reduction of the retention time. This factor is responsible for poor bioavailability of phytodrugs.

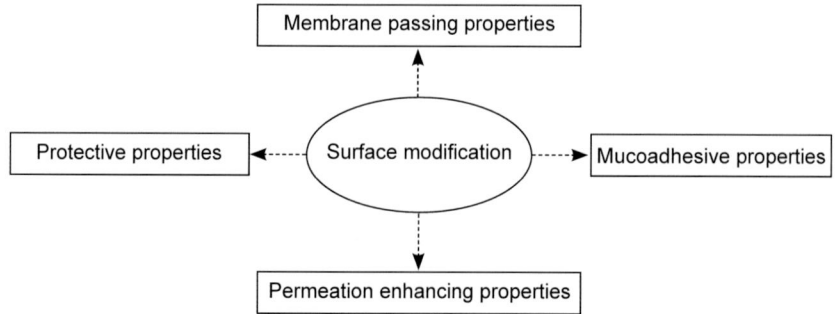

Fig. 21. Application of surface modification in drug-delivery of phytomedicine (Adapted from Gunasekaran et al. 2014).

Nanophytomedicine (incorporating phytomedicine in nanoparticles) provides best possible solution in the improvement of bioavailability and retention time of phytomedicines in the epithelial or mucosal tissue. Phytomedicines have low mucosal tissue absoption property. Nanophytomedicine, when administered by the oral route, has better diffusion in the alimentary tract. Nanophytomedicine comes in contact with the mucosal tissue and provides the necessary surface for adhesion.

4.5.3 Nano carriers for phytomedicine

The following methods have been used for formulations based on nanophytomedicine (Fig. 22).

- complex coacervation,
- co-precipitation,
- homogenization,
- nanoprecipitation or solvent displacement,
- salting out,
- self assembly,
- solvent emulsification-diffusion,
- supercritical fluid.

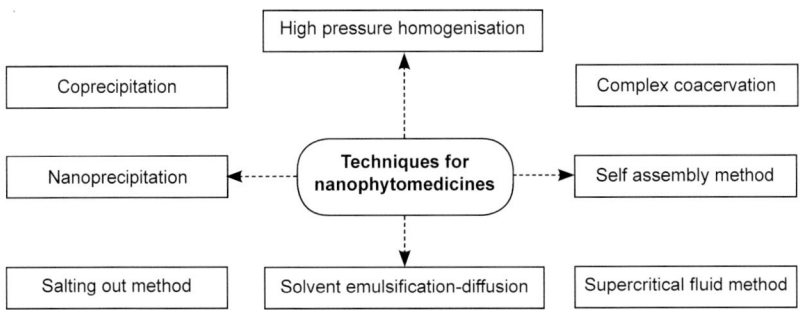

Fig. 22. It demonstrates common techniques used in nanophytomedicine (Adapted from Gunasekaran et al. 2014).

4.5.4 Nanophytopharmaceuticals

The method mentioned above can be used in preparation of different types of nanophytopharmaceuticals (Fig. 23).

- colloidal nano-liposomes,
- dendrimers,
- magnetic nanoparticles,
- metal and inorganic nanoparticles,
- phospholipid micelles,
- polymeric micelles,
- polymeric nanoparticles,

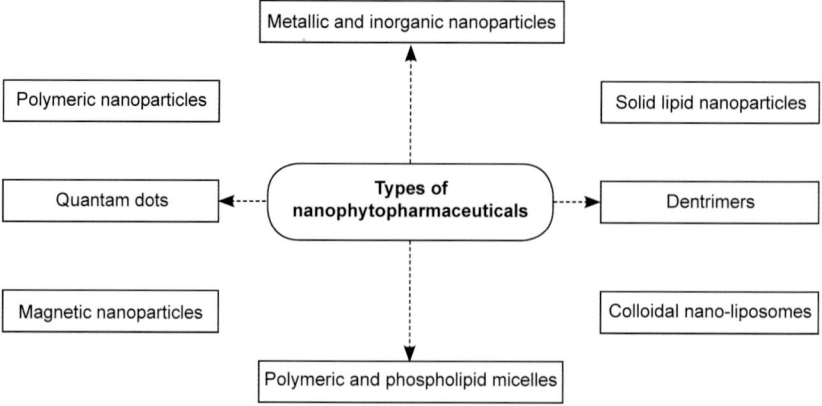

Fig. 23. Diagram showing a variety of nanophytopharmaceuticals (Adapted from Gunasekaran et al. 2014).

- quantum dots,
- solid lipid nanoparticles.

4.6 Nanophytomedicine and Ayurvedic Formulations

Nanophytomedicine is a newly emerging area in the field of Ayurveda, the traditional Indian Medicine. Nanophytomedicine allows encapsulation of the active constituents of medicinal plants for the treatment of an array of diseases ranging from infection to diabetes.

Ayurvedic nanocapsules are the herbal drug containing nanoshells made with a polymers of non-toxic origin. The nanocapsules are employed for the purpose of controlled drug delivery at a specific site in a targeted fashion. The type of polymers used for herbal nanocapsules preparations are Poly-e-caprolactone (Fig. 24), polylactide (Fig. 25), and poly(D,L-lactide-co-glycolide) (Fig. 26).

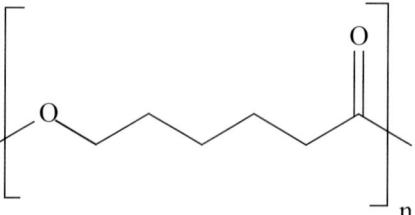

Fig. 24. Structure of Poly-e-caprolactone.

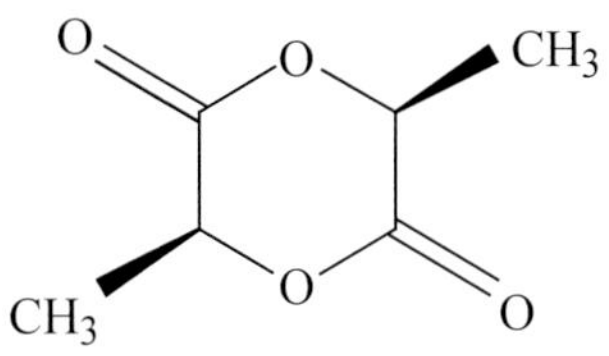

Fig. 25. Structure of Polylactide.

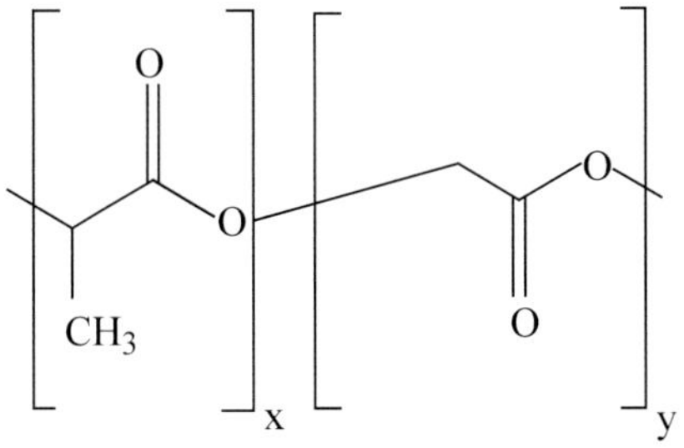

Fig. 26. Structure of Poly(D,L-lactide-co-glycolide).

4.7 Nanocapsules in Phytomedicine

The application of nanocapsules in the field of phytomedicine is due to their small size and high surface area to volume ratio. This further, improves pharmacokinetic

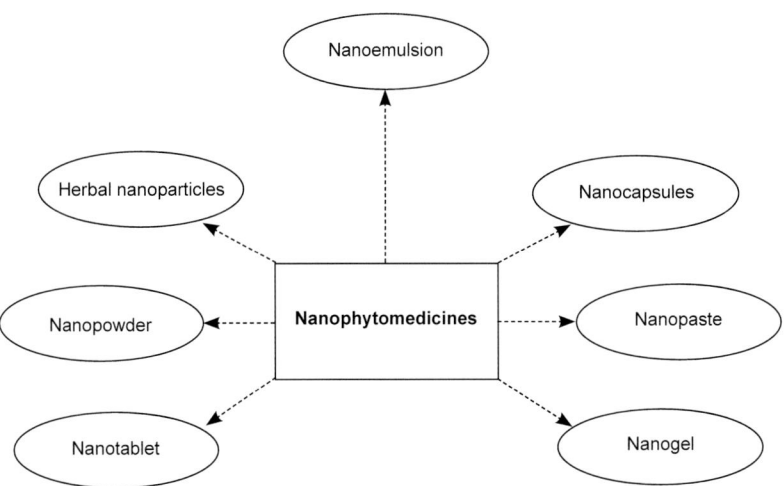

Fig. 27. Nanophytomedical Drug Formulations (Adapted from Othayoth et al. 2014).

profile and biodistribution of the therapeutic agent by nanoparticles drug carriers. They can cross blood barrier and improve the solubility and stability of hydrophobic compounds (Othayoth et al. 2014).

Further Reading

Ajazuddin SS. Applications of novel drug delivery system for herbal formulations. *Fitoterapia.* 2010; **81**: 680–689.

Ansari SH, Islam F, Mohd S. Influence of nanotechnology on herbal drugs: A Review. *Journal of Advanced Pharmaceutical Technology & Research.* 2012; **3**: 142–6.

Bell IR, Sarter B, Standish LJ, Banerji P, Banerji P. Low Doses of Traditional Nanophytomedicines for Clinical Treatment: Manufacturing Processes and Nonlinear Response Patterns. *J Nanosci Nanotechnol.* 2015; **15**: 4021–38.

Gunasekaran T, Haile T, Nigusse T, Dhanaraju MD. Nanotechnology: an effective tool for enhancing bioavailability and bioactivity of phytomedicine. *Asian Pac J Trop Biomed.* 2014; **4**: 1–7.

Othayoth R, Kalivarapu S, Botlagunta M. Nanophytomedicine and drug formulations. *Int J Nanotechnol Appl.* 2014; **4**: 1–8.

Patravale VB, Fernandes C, Pol A, Patel P, Parekh V. A Special Section on Nanophytomedicine. *J Nanosci Nanotechnol.* 2015; **15**: 4019–20.

CHAPTER 5

Nanocurcumin

5.1 What is Nanocurcumin?

The poor rate of absorption and in the intestine coupled with rapid elimination from the body are the main contributing factors for the poor bioavailability of curcumin. Solubility in the aqueous medium and permeability through the gut are the two factors contributing to the absorption of curcumin. The extracts of *Curcuma longa* standardised to curcumin are lacking both of these factors marketed as capsules and tablets.

A variety of nano-preparations of curcumin, including curcumin conjugates, cyclodextrin complexes, liposomes, micelles, nanodisks, nanofibres, nanoemulsions, nanoparticles, and solid lipid nanoparticles have been introduced for the treatment of carcinoma (Liu et al. 2013).

In recent years, the researchers have stressed on improving the kinetics of curcumin. In recent years, the application of nanotechnology has contributed tremendously in improving the therapeutic efficacy of curcumin. The formulations of curcumin based on nanotechnology are known as "nanocurcumin" (Flora et al. 2013).

5.2 Synthesis of Nanocurcumin

5.2.1 Copolymers technique

The synthesis of nanocurcumin has been achieved by the utilization of the micellar aggregates of cross-linked and random copolymers of N-isopropylacrylamide (Fig. 28), with N-vinyl-2-pyrrolidone (Fig. 29), and poly(ethyleneglycol)monoacrylate (Fig. 30) (Bisht et al. 2007). Dynamic laser light scattering and transmission electron microscopy have confirmed a narrow size distribution in the 50 nm range (Bisht et al. 2007).

5.2.2 Wet-milling technique

A process utilising wet-milling technique has been employed in the preparation of nanoparticles of curcumin. Nanocurcumin has been found to have a narrow particle size distribution in the range of 2–40 nm (Bhawana et al. 2011).

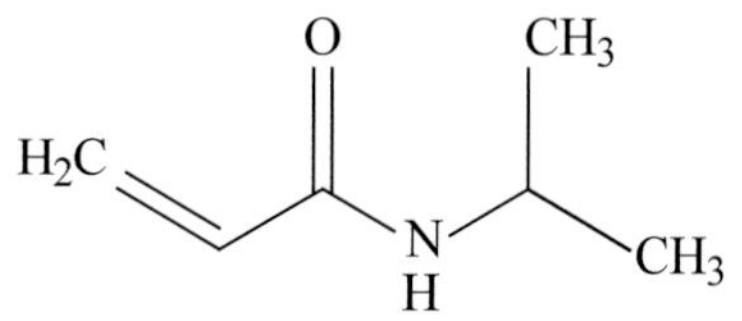

Fig. 28. Structure of *N*-isopropylacrylamide.

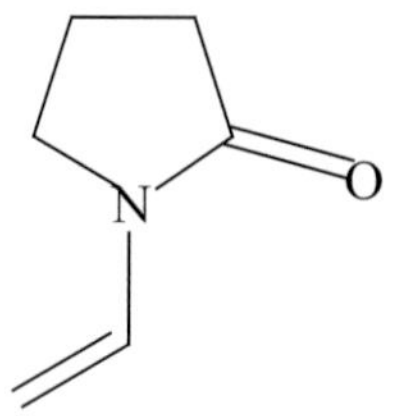

Fig. 29. Structure of *N*-vinyl-2-pyrrolidone.

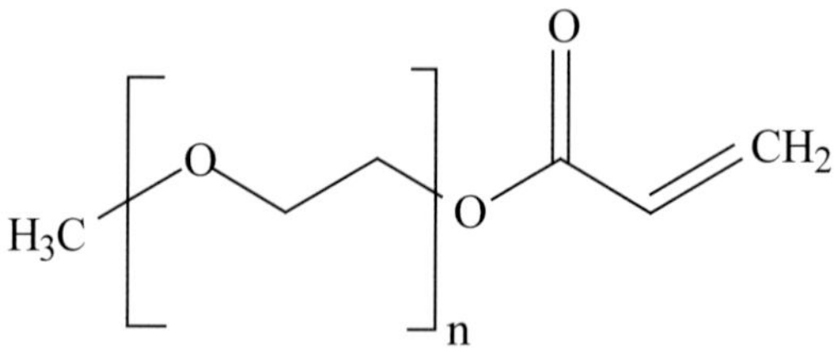

Fig. 30. Structure of poly(ethyleneglycol) monoacrylate.

5.2.3 Nanoparticle technology

'Submicron curcumin' has been formed with a yield of 92.5% (average diameter of 546 nm). The formation has been achieved without the use of added additive. Centri fugation helped in reducing the average of 'submicron curcumin' to 364 nm. The technique of high pressure homogenization in association with a food matrix decreased the average particle size to 99 nm in the form of liposomes (Chinh 2009).

5.2.4 Femtosecond laser ablation

During the pulverisation stage in femtosecond laser ablation, the average size of curcumin was affected by laser scan speed and strength and buffer solution. The minimum particle size was 500 nm and was found to be stable after a period of 30 days (Tagami et al. 2014).

5.2.5 Double emulsion technique

The nanoparticles were prepared by a combination of poloxamers and poly(lactic-*co*-glycolic acid). The interaction of the nanoparticles with the serum proteins confirmed the distinct surface properties of the nanoparticles prepared from poly(lactic-*co*-glycolic acid) alone and a combination of poloxamers and poly(lactic-*co*-glycolic acid) (Mayol et al. 2015).

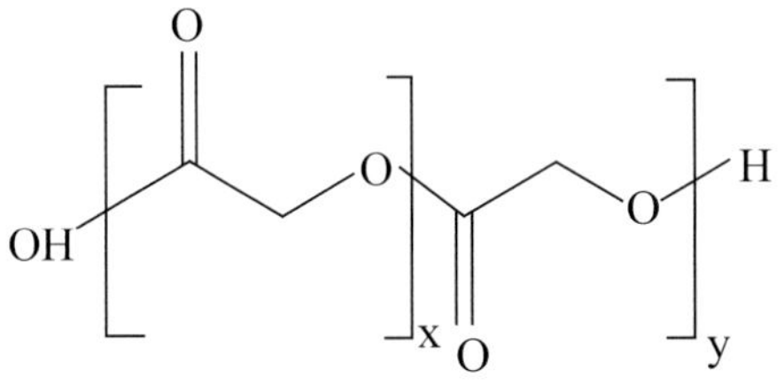

Fig. 31. Structure of poly(lactic-*co*-glycolic acid).

5.3 Pre-clinical Pharmacology

5.3.1 Antimicrobial

5.3.1.1 Antibacterial

The aqueous dispersion of nanocurcumin was found to be more effective than curcumin against bacteria (*Bacillus subtilis, Escherichia coli, Pseudomonas aeruginosa,* and *Staphylococcus aureus*) and fungi (*Aspergillus niger* and *Penicillium notatum*). The reduction of the particle size up to a nanorange resulted in an improvement in antimicrobial activity and water solubility of curcumin (Bhawana et al. 2011).

A study investigated the antibacterial effect of microcurcumin, macrocurcumin and nanocurcumin against predominant pathogenic bacteria. Nanocurcumin demonstrated a fivefold enhancement in bactericidal activity (Gopal et al. 2016).

5.3.1.2 Antifilarial

Nanocurcumin in a dose of 5×5 mg/kg, orally significantly augmented the adulticidal microfilariciadal and microfilariciadal action of curcumin in comparison to free curcumin in a dose of 5×50 mg/kg, orally or diethylcarbamizine in a dose of 50 mg/kg, orally (Fig. 32), the standard antifilarial agent against the Brugia malayi Mastomys coucha rodent model. Absolute elimination of wolbachia from the recovered female parasites was proved by the polymerase chain reaction (Ali et al. 2014).

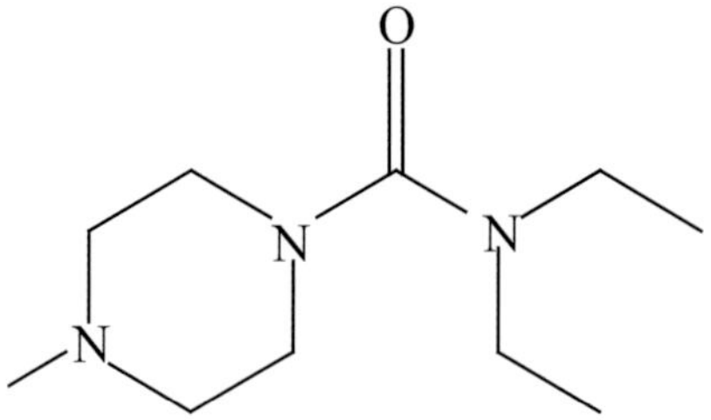

Fig. 32. Structure of diethylcarbamizine, an antifilarial agent.

5.3.1.3 Antileishmanial

A study investigated the possible role of nanocurcumin as a possible antileishmanial agent in animal model of visceral leishmaniasis. The effect was seen on enlarged spleen (splenomegaly) and delayed hypersensitivity. Nanocurcumin demonstrated significant efficacy *in vivo* in comparison to free curcumin and pentamidine (Samim et al. 2011).

5.3.2 Anticancer

5.3.2.1 Esophageal adenocarcinoma

A study investigated the effectiveness of nanocurcumin against esophageal adenocarcinoma cells. Nanocurcumin significantly resulted in decrease of the proliferation of the esophageal adenocarcinoma cells. T cells were combined with nanocurcumin in OE19 and OE33 cells, the basic levels of T cell induced cytotoxicity of 6.4 and 4.1%, increased to 15 and 13%, respectively (Milano et al. 2013).

5.3.2.2 Hepatocellular carcinoma

In vitro and *in vivo* studies evaluated the efficacy of polymeric nanoparticle formulation of curcumin alone and in combination with sorafenib (Fig. 33) in human hepatocellular carcinoma. *In vitro*, nanocurcumin inhibited the proliferation and invasion of human hepatocellular carcinoma cell lines. Further, *in vivo*, nanocurcumin significantly caused suppression of the primary tumor growth and lung metastases. The combination of nanocurcumin and sorafenib induced cell apoptosis and cell cycle arrest (Hu et al. 2015).

A study investigated the impact of dendrosomal nanocurcumin on expression of mir-34 family members in two HCC cell lines (HepG2 and Huh7). Gene expression assays showed that dendrosomal nanocurcumin upregulated mir34a, mir34b and mir34c expression ($P < 0.05$) as well as downregulated DNMT1, DNMT3A and DNMT3B expression ($P < 0.05$) in both cell lines (Chamani et al. 2016).

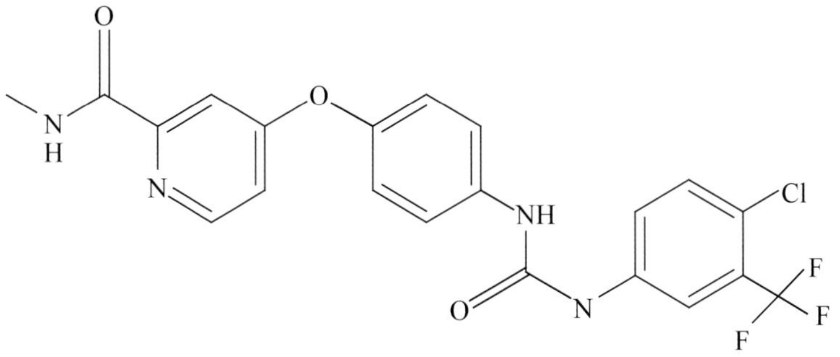

Fig. 33. Structure of sorafenib (a kinase inhibitor for treatment of human hepatocellular carcinoma).

5.3.2.3 Colon cancer

A study investigated the effects of polymeric nanocarrier-curcumin on colon cancer in an azoxymethane-induced rat tumor. Forty rats were divided into control, curcumin and polymeric nanocarrier-curcumin-treated groups. Animals received azoxymethane (Fig. 34) as a carcinogenic agent (15 mg/kg, s.c.) weekly for two consecutive weeks. They were given curcumin 0.2% and polymeric nanocarrier-curcumin two weeks before till 14 weeks after the last injection of azoxymethane.

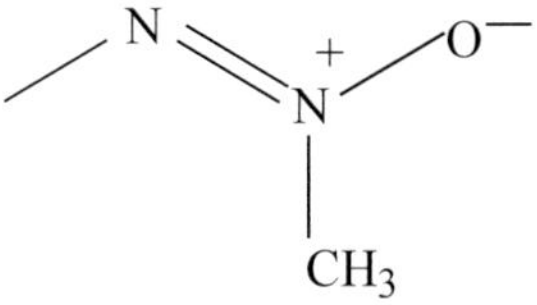

Fig. 34. Structure of azoxymethane, a carcinogenic agent.

In vivo, curcumin nanoparticles inhibited colon cancer growth in animal model. The tumors incidence and number decreased by nanocurcumin comparison with control (Alizadeh et al. 2012).

5.3.2.4 Human pancreatic cancer

Nanocurcumin demonstrated comparable *in vitro* efficacy to free curcumin against human pancreatic cancer cell lines. The results were assessed by cell viability and clonogenicity assays in soft agar (Bisht et al. 2007).

5.3.2.5 Medulloblastoma and glioblastoma

Nanocurcumin caused a dose-dependent decrease in cell growth as measured by MTT in multiple brain cancer cell lines. Nanocurcumin reduced the CD133+ stem-like cancer cell population in medulloblastoma and glioblastoma cells (Lim et al. 2010).

5.3.2.6 Nasopharyngeal cancer

With the help of atomic force microscopy detailed characterization of curcumin-cell interaction was carried out in human nasopharyngeal cancer cells. Nanocurcumin showed an enhanced uptake over micron sized drugs (Prasanth et al. 2011).

5.4 Antioxidant

A Phase 2 study investigated the comparative antioxidant effect of curcumin nanocrystals and bulk curcumin in rats treated with 1,2-dimethyl hydrazine. The decrease in lipid peroxidation was the mode of action behind the antioxidant effect of curcumin nanocrystals. The exhibition of the antioxidant effect of curcumin nanocrystals was at a does of 40 mg. On the other hand, the bulk curcumin exerted the antioxidant effect at a dose of 80 mg (Rajasekar and Devasena 2015).

5.5 Antigenotoxicity

In this study, group I rats were taken as the control group. The rats in group II were fed with sodium arsenite ($NaAsO_2$) in a dose of 25 ppm on a daily basis in drinking water for fourteen days. The maintenance of the rats in group III, IV and V was similar to the group II. The rats in group III, IV and V were fed with empty nanoparticle, curcumin and curcumin loaded nanoparticles, respectively. Dose of both curcumin and curcumin loaded nanoparticles was 100 mg/kg bw. Both forms of curcumin reduced the genotoxicity caused by arsenic (Sankar et al. 2014).

5.6 Hepatoprotective

In yet another study, group I rats were taken as the control group. The rats in group II were fed with sodium arsenite ($NaAsO_2$) in a dose of 25 ppm on a daily basis in drinking water for fourty-two days. The maintenance of the rats in group III, IV and V was similar to the group II. The rats in group III, IV and V were fed with empty nanoparticle, curcumin and curcumin loaded nanoparticles, respectively. Dose of both curcumin and curcumin loaded nanoparticles was 100 mg/kg bw. Both forms of curcumin reduced the hepatotoxicity caused by arsenic (Sankar et al. 2014).

5.7 Immunomodulatory

In yet another study, a similar schedule was followed as in antigenotoxicity and hepatoprotective studies. Arsenic resulted in significant decrease in the proliferation of lymphocytes in the spleen in response to the Keyhole Limpet Hemocyanin (metalloprotein) and mitogen-concanavalin-A (lectin). Free curcumin and nanocurcumin treatment significantly attenuated effects mediated through arsenic (Sankar et al. 2013).

5.8 Pharmacokinetics

A comparative study explored ADME of nanocurcumin and solvent-solubilized curcumin in rats. Both forms of curcumin were administered to rats in a dose of 10 mg/kg. Nanocurcumin caused increase in the plasma Cmax of curcumin manifold as compared to solvent-solubilized curcumin. Nanocurcumin also resulted in increased relative availability of curcumin and metabolites (Zou et al. 2013).

Further Reading

Ali M, Afzal M, Abdul Nasim S, Ahmad I. Nanocurcumin: a novel antifilarial agent with DNA topoisomerase II inhibitory activity. *J Drug Target.* 2014; **22:** 395–407.

Bhawana N, Basniwal RK, Buttar HS, Jain VK, Jain N. Curcumin nanoparticles: preparation, characterization, and antimicrobial study. *J Agric Food Chem.* 2011; **59:** 2056–61.

Bisht S, Feldmann G, Soni S, Ravi R, Karikar C, Maitra A, Maitra A. Polymeric nanoparticle-encapsulated curcumin ("nanocurcumin"): a novel strategy for human cancer therapy. *J Nanobiotechnol.* 2007; **5:** 3.

Chamani F, Sadeghizadeh M, Masoumi M, Babashah S. Evaluation of MiR-34 Family and DNA Methyltransferases 1, 3A, 3B Gene expression levels in hepatocellular carcinoma following treatment with dendrosomal nanocurcumin. *Asian Pac J Cancer Prev.* 2016; **17:** 219–24.

Chinh VN. Preparation of "Submicron Curcumin" and "Nanocurcumin" from *Curcuma longa* L. Using Nanoparticle Technology. 2009; https://repository.ugm.ac.id/95282.

Flora G, Gupta D, Tiwari A. Nanocurcumin: a promising therapeutic advancement over native curcumin. *Crit Rev Ther Drug Carrier Syst.* 2013; **30:** 331–68.

Gopal J, Muthu M, Chun S. Bactericidal property of macro-, micro- and nanocurcumin: An assessment. *Arab J Sci Eng.* 2016; **41:** 2087–2093.

Hu B, Sun D, Sun C, Sun YF, Sun HX, Zhu QF, Yang XR, Gao YB, Tang WG, Fan J, Maitra A, Anders RA, Xu Y. A polymeric nanoparticle formulation of curcumin in combination with sorafenib synergistically inhibits tumour growth and metastasis in an orthotopic model of human hepatocellular carcinoma. *Biochem Biophys Res Commun.* 2015; **468:** 525–32.

Lim KJ, Maitra A, Bisht S, Eberhart C, Bar E. In Proceedings of the 101st Annual Meeting of the American Association for Cancer Research; 2010 Apr 17–21; Washington, DC. Philadelphia (PA): AACR; *Cancer Res.* 2010; **70:** Abstract no 4440.

Liu J, Chen S, Lv L, Song L, Guo S, Huang S. Recent progress in studying curcumin and its nano-preparations for cancer therapy. *Curr Pharm Des.* 2013; **19:** 1974–93.

Mayol L, Serri C, Menale C, Crispi S, Piccolo MT, Mita L, Giarra S, Forte M, Saija A, Biondi M, Mita DG. Curcumin loaded PLGA-poloxamer blend nanoparticles induce cell cycle arrest in mesothelioma cells. *Eur J Pharm Biopharm.* 2015; **93:** 37–45.

Milano F, Mari L, van de Luijtgaarden W, Parikh K, Calpe S, Krishnadath KK. Nano-curcumin inhibits proliferation of esophageal adenocarcinoma cells and enhances the T cell mediated immune response. *Front Oncol.* 2013; **3:** 137.

Prasanth R, Nair G, Girish C. Enhanced endocytosis of nano-curcumin in nasopharyngeal cancer cells: An atomic force microscopy study. *Appl Physics Lett.* 2011; 99.

Rajasekar A, Devasena T. Facile Synthesis of Curcumin Nanocrystals and Validation of Its Antioxidant Activity Against Circulatory Toxicity in Wistar Rats. *J Nanosci Nanotechnol.* 2015; **15:** 4119–25.

Samim M, Naqvi S, Arora I, Ahmad FJ, Maitra A. Antileishmanial activity of nanocurcumin. *Ther Deliv.* 2011; **2:** 223–30.

Sankar P, Gopal Telang AG, Kalaivanan R, Karunakaran V, Manikam K, Sarkar SN. Effects of nanoparticle-encapsulated curcumin on arsenic-induced liver toxicity in rats. *Environ Toxicol.* 2015; **30:** 628–37.

Sankar P, Telang AG, Ramya K, Vijayakaran K, Kesavan M, Sarkar SN. Protective action of curcumin and nano-curcumin against arsenic-induced genotoxicity in rats *in vivo. Mol Biol Rep.* 2014; **41:** 7413–22.

Sankar P, Telang AG, Suresh S, Kesavan M, Kannan K, Kalaivanan R, Sarkar SN. Immunomodulatory effects of nanocurcumin in arsenic-exposed rats. *Int Immunopharmacol.* 2013; **17:** 65–70.

Tagami T, Imao Y, Ito S, Nakada A, Ozeki T. Simple and effective preparation of nano-pulverized curcumin by femtosecond laser ablation and the cytotoxic effect on C6 rat glioma cells *in vitro. Int J Pharm.* 2014; **468:** 91–6.

Zou P, Helson L, Maitra A, Stern ST, McNeil SE. Polymeric curcumin nanoparticle pharmacokinetics and metabolism in bile duct cannulated rats. *Mol Pharm.* 2013; **10:** 1977–87.

CHAPTER 6

Nanoandrographolide

6.1 Pharmacology

6.1.1 Anticancer

Nanoparticles of andrographolide having coating of chitosan caused accentuation of localisation of the cells. Further, they caused induction of G1 cell cycle and resulted in increase in toxicity in the cells. The nanoparticles of andrographolide caused significant reduction in the tumour incidence (68.21%) as compared to andrographolide (24/7%). The nanoparticles of andrographolide also increased life span of the rats significantly as compared to andrographolide (Roy et al. 2012).

Another study exploring anticancer potential of andrographolide proved that nanoandropgarpholide has better bioavailability as evident from an improved Cmax and area under the curve (AUC) for later. Nanoandropgarpholide is also released in a sustained fashion as compared to andrographolide. Nanoandrographolide demonstrated significant antitumour activity as compared to andrographolide in Balb/c mice (Parveen et al. 2014).

6.1.2 Antileishmanial

Antileishmanial activity of andrographolide nanoparticles loaded in 50:50 poly(DL-lactide-co-glycolic acid) was found to be significantly greater in about 1/4th of the dosage andrographolide (IC_{50}) 160 µM (Roy et al. 2010).

6.1.3 Hepatoprotective

In an experimental study, the new functionalized andrographolide-loaded polylactide co-glycolide nanoparticles demonstrated efficient hepatoprotective activity against over-dosage of antipyretic, paracetamol (Roy et al. 2013).

A study investigated possible hepatoprotective activity of polylactide co-glycolide nanocapsulated andrographolide against hepatotoxicity caused by arsenic. Andrographolide and polylactide co-glycolide nanocapsulated andrographolide provided hepatoprotection against arsenic induced hepatotoxicity by decreasing the raised liver function tests and deposition of the hepatotoxin in the liver (Das et al. 2015).

6.1.4 Antihyperlipidemic

Nanoandrographolide demonstrated significant antihyperlipidemic activity and bioavailability. Both the things were achieved due to increase in the solubility and stability of andrographolide in the gut due to change in the transport mode in Caco 2 cell. Nanoandrographolide showed significant increase in the bioavailability (241%) as compared to suspension of andrographolide (Yang et al. 2013).

6.1.5 Bioavailability

Nanoprecipitation technique using cationic poly methacrylate copolymer (Fig. 35) was employed for the prepation of andrographolide-loaded pH-sensitive nanoparticles. In male rats, the bioavailability of andrographolide-loaded pH-sensitive nanoparticles and pure andrographolide was determined at a dose of 10 mg/kg. Nanoandrographolide showed significant increase in AUC0-∞ (2.2–3.2), Cmax and increased bioavailability (121.53%) as compared to andrographolide (Chellampillai and Pawar 2011).

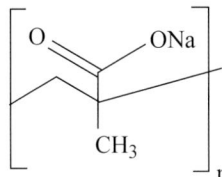

Fig. 35. Structure of poly methacrylate copolymer.

Further Reading

Chellampillai B, Pawar AP. Improved bioavailability of orally administered andrographolide from pH-sensitive nanoparticles. *Eur J Drug Metab Pharmacokinet*. 2011; **35:** 123–9.

Das S, Pradhan GK, Das S, Nath D, Das Saha K. Enhanced protective activity of nano formulated andrographolide against arsenic induced liver damage. *Chem Biol Interact*. 2015; **242:** 281–9.

Parveen R, Ahmad FJ, Iqbal Z, Samim M, Ahmad S. Solid lipid nanoparticles of anticancer drug andrographolide: formulation, *in vitro* and *in vivo* studies. *Drug Dev Ind Pharm*. 2014; **40:** 1206–12.

Roy P, Das S, Auddy RG, Saha A, Mukherjee A. Engineered andrographolide nanoparticles mitigate paracetamol hepatotoxicity in mice. *Pharm Res*. 2013; **30:** 1252–62.

Roy P, Das S, Bera T, Mondol S, Mukherjee A. Andrographolide nanoparticles in leishmaniasis: characterization and *in vitro* evaluations. *Int J Nanomedicine*. 2010; **5:** 1113–21.

Roy P, Das S, Mondal A, Chatterji U, Mukherjee A. Nanoparticle engineering enhances anticancer efficacy of andrographolide in MCF-7 cells and mice bearing EAC. *Curr Pharm Biotechnol*. 2012; **13:** 2669–81.

Yang T, Sheng HH, Feng NP, Wei H, Wang ZT, Wang CH. Preparation of andrographolide-loaded solid lipid nanoparticles and their *in vitro* and *in vivo* evaluations: characteristics, release, absorption, transports, pharmacokinetics, and antihyperlipidemic activity. *J Pharm Sci*. 2013; **102:** 4414–25.

Nanohypericum, Nanohypericin and Nanohyperforin

7.1 Pharmacology of Nanohypericin

7.1.1 Antibacterial

Hypericin-laden nanoparticles were prepared from amphiphilic copolymers and *in vitro* antimicrobial activity was determined against methicillin resistant *Staphylococcus aureus*. *In vivo* antimicrobial activity was carried out in rats on wounds having infection. The results of *in vivo* study demonstrated significant wound-healing activity of hypericin-laden nanoparticles in terms of parameters (Nafee et al. 2013).

In yet another study, synthesis of hypericin silver nanoparticles was achieved from shoot cultures of *Hypercium hookerianum* Wight & Arn. (Hooker's St. Johnswort). The hypericin silver nanoparticles showed significant antibacterial activity against *Bacillus subtillis* and *Pseudomonas aeruginosa*. The antibacterial activity of the hypericin silver nanoparticles increased with the rising concentration of the napthodianthrone (Manoj et al. 2015).

7.1.2 Anticancer

7.1.2.1 Ovarian cancer

An *in vitro* study investigated photoactivity of hypericin-laden nanoparticles on the NuTu-19 epithelial ovarian cancer cell line. The hypericin-laden nanoparticles demonstrated a higher photoactivity as compared to free hypericin. Increase in the dose of the light resulted in enhancement in the photoactivity of the hypericin-laden nanoparticles (Zeisser-Labouèbe et al. 2006).

7.2 Pharmacology of Nanohyperforin

7.2.1 Anti-inflammatory

An experimental study using an animal model of multiple sclerosis investigated the efficacy of hyperforin-laden gold nanoparticles in autoimmune encephalomyelitis.

Both, hyperforin-laden gold nanoparticles and hyperforin resulted in a significant reduction in clinical severity of experimental autoimmune encephalomyelitis. Hyperforin-laden gold nanoparticles resulted in a significant reduction of cytokines associated with experimental autoimmune encephalomyelitis as compared to free hyperforin group (Nosratabadi et al. 2016).

7.3 Pharmacology of Nanohypericum

Nanohypericum refers to *Hypericum perforatum* nanoparticles.

7.3.1 Antistress

A study investigated the antistress activity of *H. perforatum* gold nanoparticles in mice. The rats were fed with 200 mg/kg and 20 mg/kg of *H. perforatum* and gold nanoparticles. *H. perforatum* in both forms significantly reduced the stress induced behavioural and oxidative changes (Prakash et al. 2010).

7.3.2 Anticancer

An *in vitro* study evaluated the drug release profile of *H. perforatum* loaded nanoparticles and doxorubicin loaded nanoparticles. Entrapment efficiency was 48 and 21% for doxorubicin loaded nanoparticles and *H. perforatum* loaded nanoparticles, respectively. Doxorubicin proved to more toxic than *H. perforatum* as per MTT assay (Amjadi et al. 2013).

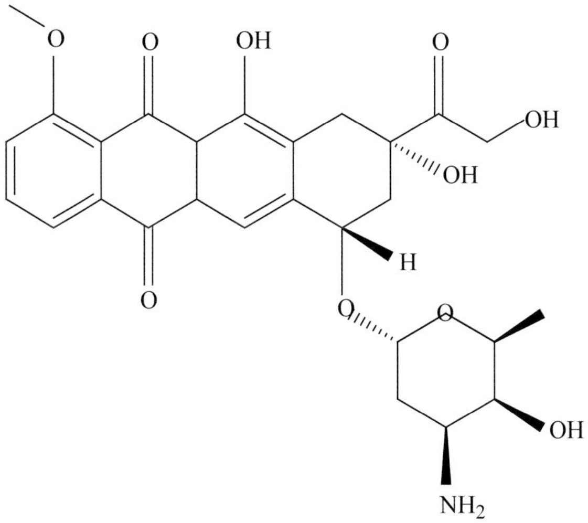

Fig. 36. Structure of doxorubicin.

Further Reading

Amjadi I, Rabiee M, Hosseini M, Mozafari M. Nanoencapsulation of *Hypericum perforatum* and doxorubicin anticancer agents in PLGA nanoparticles through double emulsion technique. *Micro & Nano Letters*. 2013; **8**: 243–247.

Manoj L, Vishwakarma V, Samal SS, Seeni S. Green synthesis of silver nanoparticles using hypericin-rich shoot cultures of *Hypericum hookerianum* and evaluation of anti-bacterial activities. *J Exp Nanosci* 2015; **10**.

Nafee N, Youssef A, El-Gowelli H, Asem H, Kandil S. Antibiotic-free nanotherapeutics: hypericin nanoparticles thereof for improved *in vitro* and *in vivo* antimicrobial photodynamic therapy and wound healing. *Int J Pharm*. 2013; **454**: 249–58.

Nosratabadi R, Rastin M, Sankian M, Haghmorad D, Mahmoudi M. Hyperforin-loaded gold nanoparticle alleviates experimental autoimmune encephalomyelitis by suppressing Th1 and Th17 cells and upregulating regulatory T cells. *Nanomedicine*. 2016; **9634**: 30028–4.

Prakash DJ, Arulkumar S, Sabesan M. Effect of nanohypericum (*Hypericum perforatum* gold nanoparticles) treatment on restraint stress induced behavioral and biochemical alteration in male albino mice. *Pharmacognosy Res*. 2010; **2**: 330–4.

Zeisser-Labouèbe M, Lange N, Gurny R, Delie F. Hypericin-loaded nanoparticles for the photodynamic treatment of ovarian cancer. *Int J Pharm*. 2006; **326**: 174–81.

CHAPTER 8

Nanotaxol

8.1 Anticancer

8.1.1 B16F10 murine melanoma

A reverse microemulsion methodology was utilised to prepare polyvinylpyrrolidone nanoparticles containing taxol. The size of the polyvinylpyrrolidone nanoparticles containing taxol was 50–60 nm. An *in vivo* study evaluated antitumour activity of encapsulated nanotaxol in B16F10 murine melanoma transplanted in C57B1/6 mice. Nanotaxol not only significantly caused reduction in the volume of the tumour but also increased the survival time as compared to free taxol (Sharma et al. 1996).

8.1.2 Breast cancer

miRi-221/222 are short RNA molecules and are hydrophilic in nature. By the process of co-precipitation, the molecules were encapsulated with calcium phosphate in a water-in-oil emulsion. The precipitates were further coated with dioleoylphosphatidic acid (Fig. 37) so as to co-encapsulate hydrophobic paclitaxel outside the hydrophilic precipitates and inside the same nanoparticle. The nanoparticle system simultaneously delivered paclitaxel and miRi-221/222 to their intracellular targets, leading to inhibit proliferative mechanisms of miR-221/222 and thus significantly enhancing the therapeutic efficacy of paclitaxel (Zhou et al. 2016).

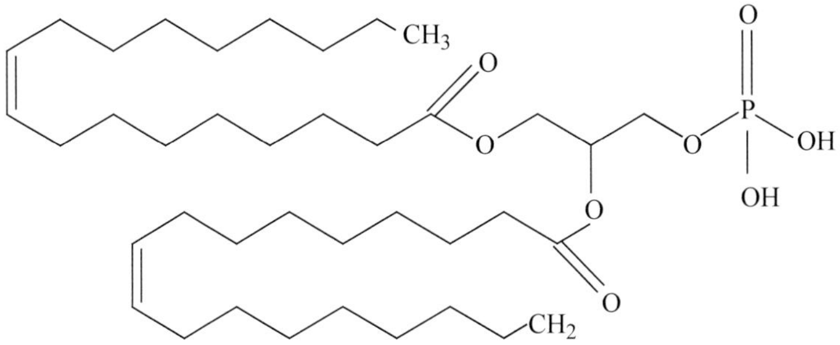

Fig. 37. Structure of dioleoylphosphatidic acid (DOPA).

The cholic acid-core polylactide-d-α-tocopheryl polyethylene glycol 1000 succinate nanoparticles demonstrated significant antitumor activity as compared to poly(lactide)-tocopheryl polyethylene glycol succinate nanoparticles and paclitaxel-loaded poly(d,l-lactide-*co*-glycolide) nanoparticles *in vitro* and *in vivo* (Tang et al. 2013).

8.1.3 Intravesical bladder cancer

The desolvation methodology was used to prepare paclitaxel loaded nanoparticles. The paclitaxel loaded nanoparticles demonstrated anticancer activity against human RT4 urinary bladder transitional cell line (Lu et al. 2004).

8.1.4 Malignant glioma

In vitro, paclitaxel nanoparticles demonstrated super paramagnetism and released paclitaxel in an extended fashion over 15 days. The paclitaxel nanoparticles easily penetrated the glioma cells and demonstrated significant toxicity in comparison to free paclitaxel (Zhao et al. 2010).

8.2 Multidrug Resistance

Ac-Arg-Val-Arg-Arg-Cys(S*t*Bu)-Lys(taxol)-2-cyanobenzothiazole (CBT-Taxol) is a taxol derivative obtained by the process of condensation. Based on the *in vitro* and *in vivo* studies, CBT-Taxol showed manifold increase in anti-multidrug resistance effects as compared with taxol on taxol-resistant HCT 116 cancer cells (Yuan et al. 2015).

8.3 Paclitaxel Nanoparticle Albumin-bound

Nanoparticle albumin-bound paclitaxel, in short known as nab-paclitaxel, is a novel solvent free form of paclitaxel (Foote 2007). The preparation of nab-paclitaxel is achieved by homogenization of paclitaxel under high pressure in the presence of serum albumin in a nanoparticle colloid system (Stinchcombe 2007). Based on the clinical evidence, nab-paclitaxel is potent than Cremophor-EL-paclitaxel (Gradishar 2006).

A phase III trial studied the efficacy of nab-paclitaxel and Cremophor-EL-paclitaxel in patients with metastatic breast carcinoma. Patients who received nab-paclitaxel reported more response and long period to progress of the cancer (Stinchcombe 2007).

Nanoparticle albumin-bound (nab)-paclitaxel has been reported to increase gemcitabine (used to treat a range of cancers) concentration inside the pancreas ductal adenocarcinoma cancer cells. nab-paclitaxel is reported to inhibit cytidine deaminase, responsible for degradation of gemcitabine (Fig. 38) (Narayanan and Weekes 2015).

A phase 2 study evaluated the efficacy and tolerability of nanoparticle albumin-bound (nab) paclitaxel in platinum-refractory urothelial cancer. Nab-paclitaxel was well tolerated in patients with pretreated advanced urothelial cancer. Nab-paclitaxel also showed an significant tumour response (Ko et al. 2013).

A study reported poor toleration of the combination of nab-paclitaxel, carboplatin (Fig. 39) and gemcitabine in high-risk patient population based on the dose and schedule. The researchers concluded that alternative combinations based on nab-paclitaxel needs to be explored as the first line treatment of urothelial cancer at advance stage (Alva et al. 2014).

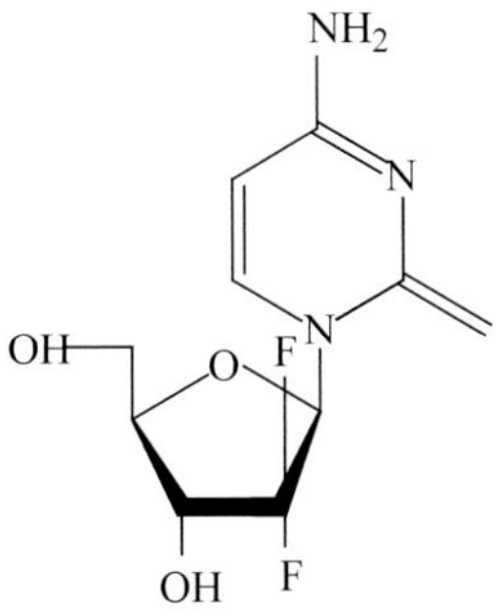

Fig. 38. Structure of gemcitabine.

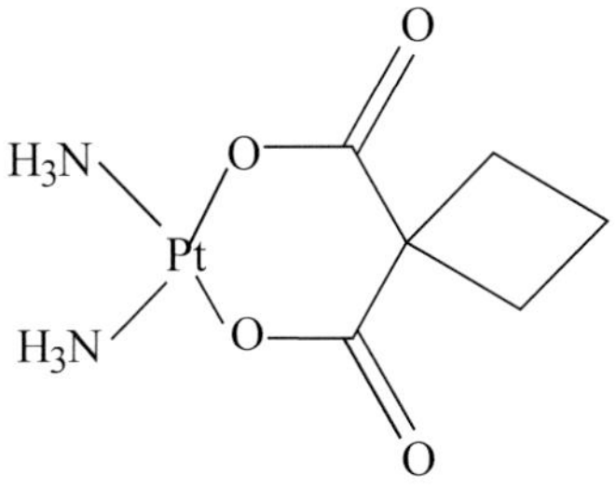

Fig. 39. Structure of carboplatin.

Further Reading

Alva A, Daignault S, Smith DC, Hussain M. Phase II trial of combination nab-paclitaxel, carboplatin and gemcitabine in first line therapy of advanced urothelial carcinoma. *Invest New Drugs*. 2014; **32:** 188–94.

Foote M. Using nanotechnology to improve the characteristics of antineoplastic drugs: improved characteristics of nab-paclitaxel compared with solvent-based paclitaxel. *Biotechnol Annu Rev.* 2007; **13:** 345–57.

Gradishar WJ. Albumin-bound paclitaxel: a next-generation taxane. *Expert Opin Pharmacother*. 2006; **7:** 1041–53.

Ko YJ, Canil CM, Mukherjee SD, Winquist E, Elser C, Eisen A, Reaume MN, Zhang L, Sridhar SS. Nanoparticle albumin-bound paclitaxel for second-line treatment of metastatic urothelial carcinoma: a single group, multicentre, phase 2 study. Lancet Oncol. 2013; **14:** 769–76.

Lu Z, Yeh TK, Tsai M, Au JL, Wientjes MG. Paclitaxel-loaded gelatin nanoparticles for intravesical bladder cancer therapy. *Clin Cancer Res*. 2004; **10:** 7677–84.

Narayanan V, Weekes CD. Nanoparticle albumin-bound (nab)-paclitaxel for the treatment of pancreas ductal adenocarcinoma. *Gastrointest Can: Targ Ther*. 2015; **5:** 11–19.

Sharma D, Chelvi TP, Kaur J, Chakravorty K, De TK, Maitra A, Ralhan R. Novel Taxol formulation: polyvinylpyrrolidone nanoparticle-encapsulated Taxol for drug delivery in cancer therapy. *Oncol Res*. 1996; **8:** 281–6.

Stinchcombe TE. Nanoparticle albumin-bound paclitaxel: a novel Cremphor-EL-free formulation of paclitaxel. *Nanomedicine* (Lond). 2007; **2:** 415–23.

Tang X, Cai S, Zhang R, Liu P, Chen H, Zheng Y, Sun L. Paclitaxel-loaded nanoparticles of star-shaped cholic acid-core PLA-TPGS copolymer for breast cancer treatment. *Nanoscale Res Lett.* 2013; **8:** 420.

Yuan Y, Wang L, Du W, Ding Z, Zhang J, Han T, An L, Zhang H, Liang G. Intracellular Self-Assembly of Taxol Nanoparticles for Overcoming Multidrug Resistance. *Angew Chem Int Ed Engl.* 2015 10; **54:** 9700–4.

Zhao M, Liang C, Li A, Chang J, Wang H, Yan R, Zhang J, Tai J. Magnetic paclitaxel nanoparticles inhibit glioma growth and improve the survival of rats bearing glioma xenografts. *Anticancer Res.* 2010; **30:** 2217–23.

Zhou Z, Kennell C, Lee JY, Leung YK, Tarapore P. Calcium Phosphate-Polymer Hybrid Nanoparticles for Enhanced Triple Negative Breast Cancer Treatment via Co-Delivery of Paclitaxel and miR-221/222 Inhibitors. *Nanomedicine.* 2016; **pii:** 1549–9634.

CHAPTER 9

Nanosilymarin

9.1 Hepatoprotective

The nanoparticles of silymarin exerted significant hepatoprotective activity against an animal model involving overdosage of paracetamol. No animal death was reported after administration of The nanoparticles of silymarin in hepatic (liver) necrosis caused by paracetamol. The regeneration of glutathione was recorded to 11.3 μmol/g in the liver tissue (Das et al. 2011).

The nanosuspension of silybin (given orally or an intravenous route) demonstrated significant hepatoprotective activity in carbon tetrachloride treated animals. The nanosuspension of silybin caused marked reduction of the raised enzymes. The histological studies also confirmed the potent hepatoprotective activity of the nanosuspension of silybin (Wang et al. 2012).

The process of nanoprecipitation was used for the preparation of PES-SIM nanoparticles and PES-SIM nanoparticles surface modified with pullulan. The nanoparticles were tested for protective activity against carbon-tetrachloride induced hepatotoxicity in rats. Pretreatment of rats PES-SIM nanoparticles and PES-SIM nanoparticles surface modified with pullulan caused in reduction the levels of enzymes as compared to carbon-tetrachloride treated group (Guhagarkar et al. 2015).

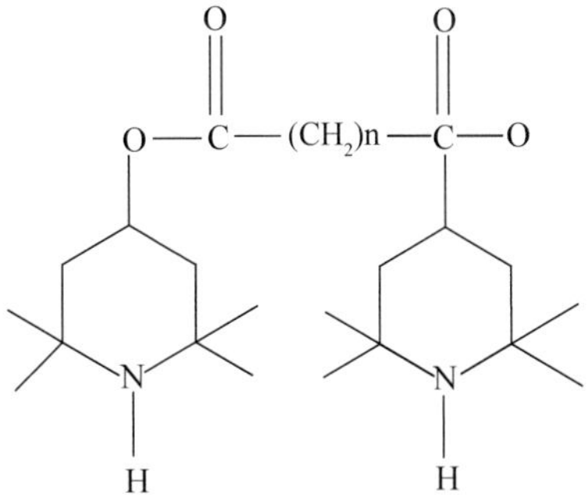

Fig. 40. Structure of polyethylene sebacate.

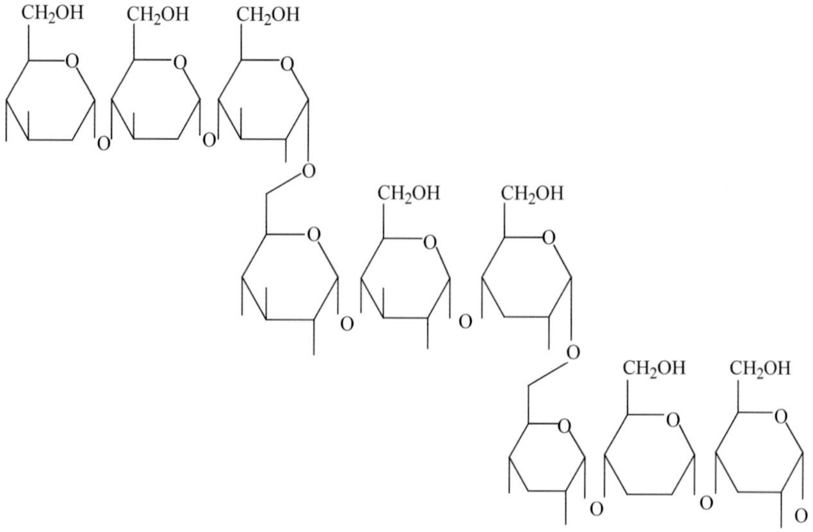

Fig. 41. Structure of pullulan.

9.2 Hepatic Fibrosis

Silymarin-loaded nanoparticles were reported to reverse the induced hepatic fibrosis. Silymarin-loaded nanoparticles significantly decreased the raised enzymes and also caused downregulation of the hepatic expression of tissue inhibitors cytokeratin-19 and metalloproteinase-1. The nanoparticles caused upregulation of the hepatic expression of matrix metalloproteinase-2. They also increase in matrix metalloproteinase-2/ metallopeptidase inhibitor-1 ratio at mRNA level (Younis et al. 2016).

9.3 A Comparative Study of Silymarin and Silymarin Nanoparticles

Nanosilymarin caused significant reduction of liver toxicity caused by D-GaIN/ tissue necrosis factor-α. The hepatoprotective activity of nanosilymarin was better as compared to silymarin (Cengiz et al. 2015).

9.4 Experimental Colitis

Trinitrobenzene sulfonic acid (Fig. 42) was the agent used for causing induction of colitis (an inflammation of the colon). The rats were divided into six groups:

Group 1: containing silymarin,
Group 2: containing nano-selenium,
Group 3: containing combination of silymarin and nano-selenium,
Group 4: containing dexamethasone (Fig. 43),
Group 5: negative control with no treatment,
Group 6: normal sham rats.

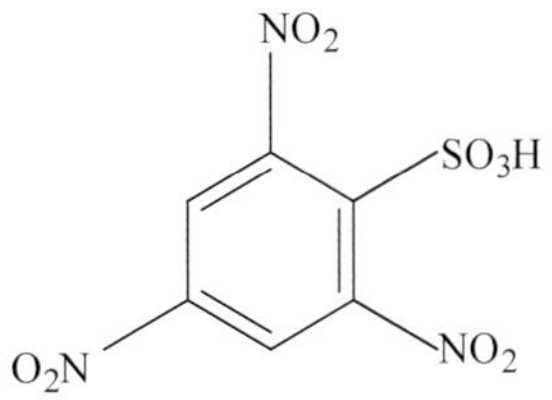

Fig. 42. Structure of trinitrobenzene sulfonic acid.

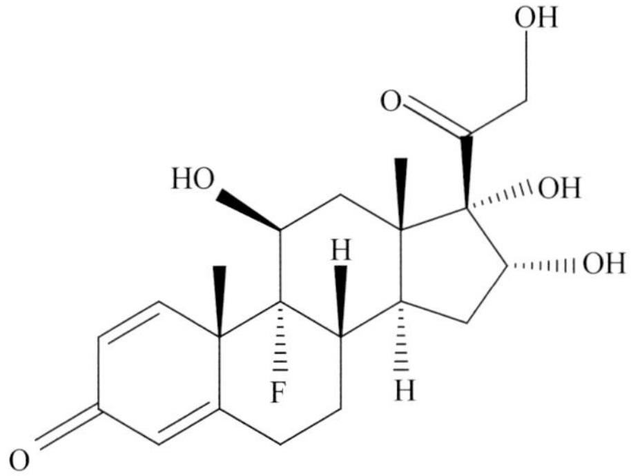

Fig. 43. Structure of dexamethasone.

The drugs were administered for a period of one week. Concomitant administration of silymarin and nano-selenium demonstrated significant antioxidant activity. They also inhibited nuclear factor-κB (Miroliaee et al. 2011).

9.5 Antioxidant

The application of nanotechnology has provided silymarin with very small size of the particle. Nanotechnology has also helped in formation of hydrogen bonding at the intermolecular level between silymarin and the matrix of the nanoparticles. Moreover, it has rendered the amorphous state of the standard hepatoprotective molecule. The above mentioned properties have contributed significantly to the dissolution profile and antioxidant activity of nanosilymarin in comparison to silymarin (Hsu et al. 2012).

9.6 Breast Cancer Metastasis

In order to block metastasis of breast cancer, nanoparticles of silibinin containing d-α tocopheryl polyethylene glycol 1000 succinate (Fig. 44) and phosphatidylcholine were synthesised. In comparison to free silibinin, silibinin-loaded lipid nanoparticles inhibited the invasion and migration of MDA-MB-231 cells. The downregulation of matrix metalloproteinase-9 and Snail is the underlying mechanism (Xu et al. 2013).

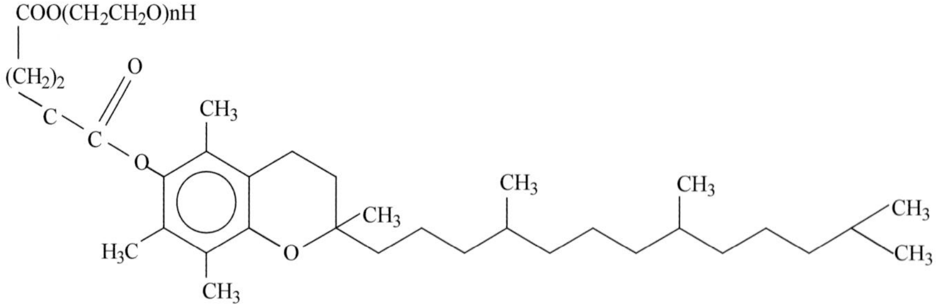

COO(CH$_2$CH$_2$O)nH

(CH$_2$)$_2$

Fig. 44. Structure of TPGS.

9.7 Prostate Cancer

In vitro cell studies showed dominant toxicity of silymarin Poly(D,L-lactic-co-glycolic acid) nanoparticles towards prostate cancer cells as compared to normal cells. Silymarin Poly(D,L-lactic-co-glycolic acid) nanoparticles showed a concentration and time dependent inhibitory effect on prostate cancer cells migration (Snima et al. 2014).

Further Reading

Cengiz M, Kutlu HM, Burukoglu DD, Ayhancı A. A comparative study on the therapeutic effects of silymarin and silymarin-loaded solid lipid nanoparticles on D-GaIN/TNF-α-induced liver damage in Balb/c mice. *Food Chem Toxicol*. 2015; **77**: 93–100.

Das S, Roy P, Auddy RG, Mukherjee A. Silymarin nanoparticle prevents paracetamol-induced hepatotoxicity. *Int J Nanomedicine*. 2011; **6**: 1291–301.

Guhagarkar SA, Shah D, Patel MD, Sathaye SS, Devarajan PV. Polyethylene Sebacate-Silymarin Nanoparticles with Enhanced Hepatoprotective Activity. *J Nanosci Nanotechnol*. 2015; **15**: 4090–3.

Hsu WC, Ng LT, Wu TH, Lin LT, Yen FL, Lin CC. Characteristics and antioxidant activities of silymarin nanoparticles. *J Nanosci Nanotechnol*. 2012; **12**: 2022–7.

Miroliaee AE, Esmaily H, Vaziri-Bami A, Baeeri M, Shahverdi AR, Abdollahi M. Amelioration of experimental colitis by a novel nanoselenium-silymarin mixture. *Toxicol Mech Methods*. 2011; **21**: 200–8.

Snima KS, Arunkumar P, Jayakumar R, Lakshmanan VK. Silymarin encapsulated poly(D,L-lactic-co-glycolic acid) nanoparticles: a prospective candidate for prostate cancer therapy. *J Biomed Nanotechnol*. 2014; **10**: 559–70.

Wang Y, Wang L, Liu Z, Zhang D, Zhang Q. *In vivo* evaluation of silybin nanosuspensions targeting liver. *J Biomed Nanotechnol*. 2012; **8**: 760–9.

Xu P, Yin Q, Shen J, Chen L, Yu H, Zhang Z, Li Y. Synergistic inhibition of breast cancer metastasis by silibinin-loaded lipid nanoparticles containing TPGS. *Int J Pharm*. 2013; **454**: 21–30.

Younis N, Shaheen MA, Abdallah MH. Silymarin-loaded Eudragit(®) Rs100 nanoparticles improved the ability of silymarin to resolve hepatic fibrosis in bile duct ligated rats. *Biomed Pharmacother*. 2016; **81**: 93–103.

CHAPTER 10

Nanoparticles of Other Bioactive Compounds

10.1 Naringenin

A naringenin loaded nanoparticles have been resolve to address the bioavailability issues of the flavonoid. Another issue was to enhance the liver protecting activity of naringenin *in vivo* when administered by the oral route. The naringenin loaded nanoparticles have been shown to have high release rate in comparison to naringenin. The naringenin loaded nanoparticles have marked improvement in solubility as compared to naringenin. Finally, the naringenin loaded nanoparticles have significant hepatoprotective activity as compared to naringenin (Yen et al. 2009).

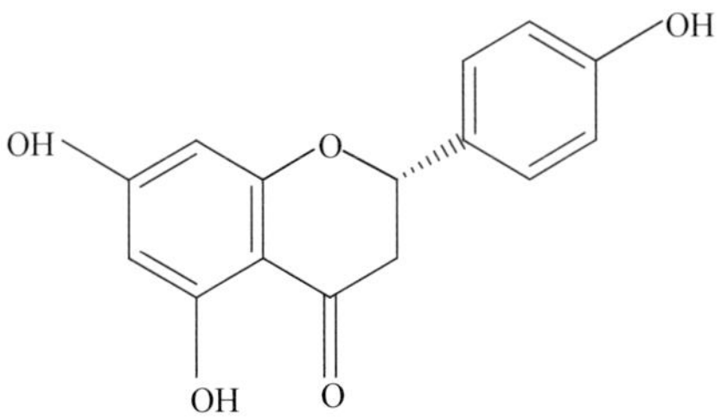

Fig. 45. Structure of naringenin.

10.2 Apigenin

The combination technology was used for the preparation of nanosuspension of apigenin. The particle size achieved was a diameter 99% (d(v)99%) of 0.515 μm. The free radical scavenging activity of the nanosuspension of flavonoid was double in comparison to the coarse suspension (Al Shaal et al. 2011).

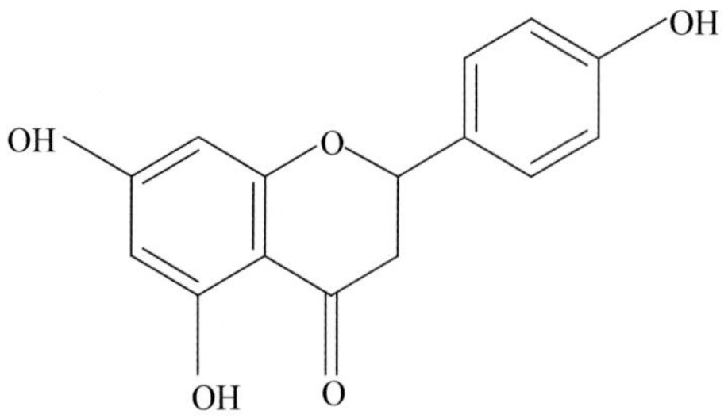

Fig. 46. Structure of apigenin.

10.3 Lutein

The nanosuspension of lutein was changed into pellets and filled in capsules of hard gelatin. This form of the flavonoid showed better release *in vitro*. Lyophilized nanosuspension of lutein was prepared for inclusion in ointment and gel. The lyophilized nanosuspension showed better re-dispersibility. The use of cellulose nitrate membranes *in vitro* system demonstrated that permeation of nanocrystals of the flavonoid was better as compared to free form (Mitri et al. 2011).

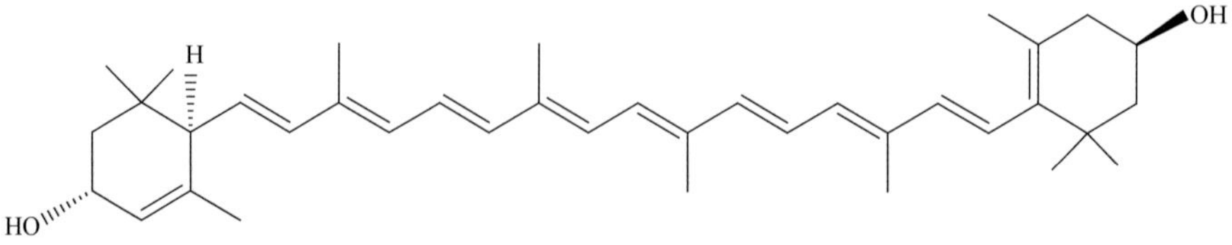

Fig. 47. Structure of lutein.

10.4 Forskolin

Coleus forskohlii roots extract was mixed with lunar caustic (sliver nitrate) and subjected to incubation. The synthesis of nanoparticles was studied with the help of ultraviolet–visible spectroscopy. The nanoparticles were found to be needle like in shape. *C. forskohlii* root extract served as a good bioreductant for the synthesis of nanoparticles. The nanoparticles showed broad spectrum antibacterial and antifungal activities (Baskaran 2013).

10.5 Naringin

Three flavonoids (hesperidin, naringin and diosmin) were used for the synthesis of silver nanoparticles. Bactericidal and cytotoxicity of the silver nanoparticles of hesperidin, naringin and diosmin was tested in human promyelocytic leukemia. Out

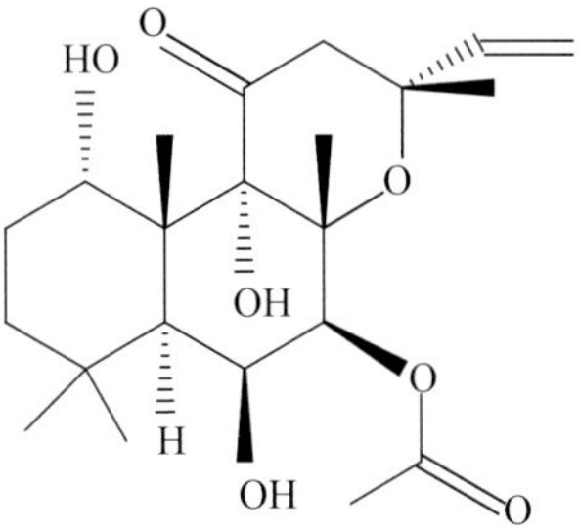

Fig. 48. Structure of forskolin.

of the three flavonoids tested, the silver nanoparticles synthesized from naringin demonstrated maximum stability and significant antibacterial and cytotoxic activities (Sahu et al. 2016).

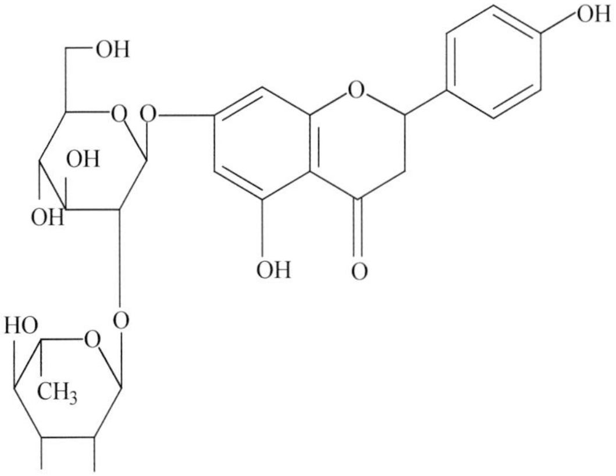

Fig. 49. Structure of naringin.

10.6 Glabridin

Nanosuspension exhibited significantly enhanced drug permeation flux of glabridin through rat skin with no lag phase both *in vitro* and *in vivo*, in comparison to the coarse suspension and physical mixture. The glabridin nanosuspension showed no significant particle aggregates and a drug loss of 5.46% after storage for 3 months at room temperature (Wang et al. 2016).

10.7 Quercetin

Nanoquercetin effectively restored integrity of the liver membrane in a animal model of carbon tetrachloride-mediated liver cirrhosis. Significant reduction of serum markers was noticed in comparison to free quercetin. The reduced collagen and histopathological findings proved that the nanoquercetin offered much better effects in comparison to free quercetin (Verma et al. 2016).

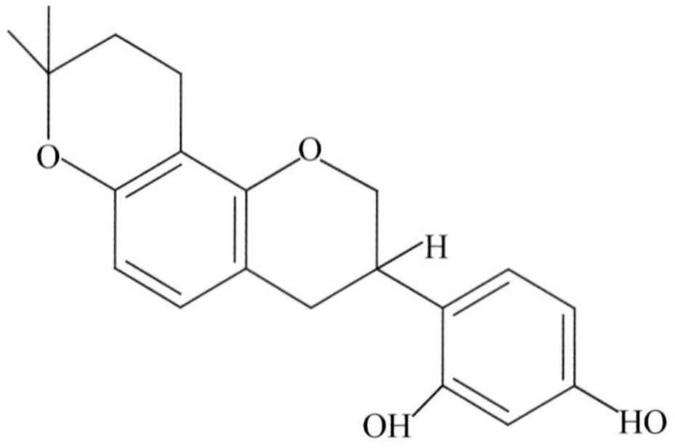

Fig. 50. Structure of glabridin.

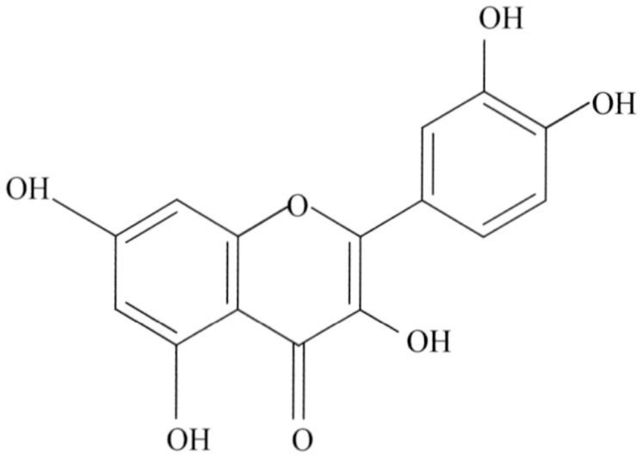

Fig. 51. Structure of quercetin.

10.8 Epigallocatechin-3-Gallate

In a mouse epidermal cell line, the inhibitory effects of epigallocatechin-3-gallate on the toxicity induced by nickel nanoparticles was investigated. The compound inhibited the toxicity to a certain extent and reduced the apoptotic cell number. The regulation of the protein changes in mitogen-activated protein kinase signalling pathway seems to be mode of action in alleviation of the nickel nanoparticles toxicity (Gu et al. 2016).

10.9 Tetrandrine

The synthesis of solid lipid nanoparticles of the alkaloid has been achieved to address the solubility issues (Li et al. 2006).

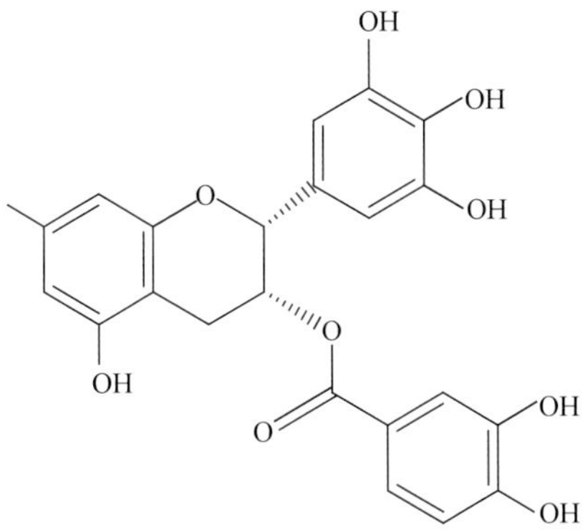

Fig. 52. Structure of epigallocatechin-3-gallate.

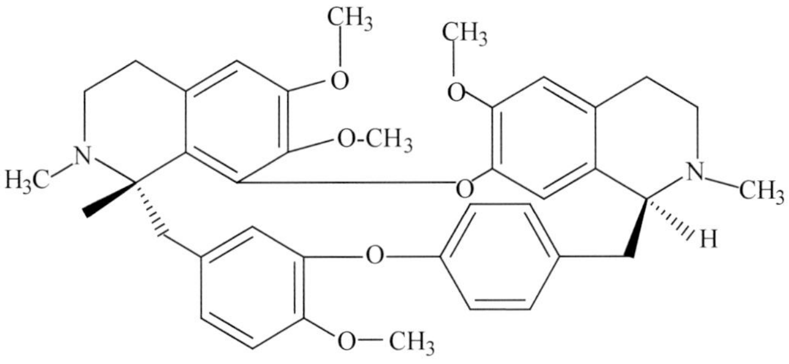

Fig. 53. Structure of tetrandrine.

10.10 Crytotanshinone

The synthesis of solid lipid nanoparticles of the phenanthrene quinone has been achieved to address the poor bioavailability issues (Hu et al. 2010).

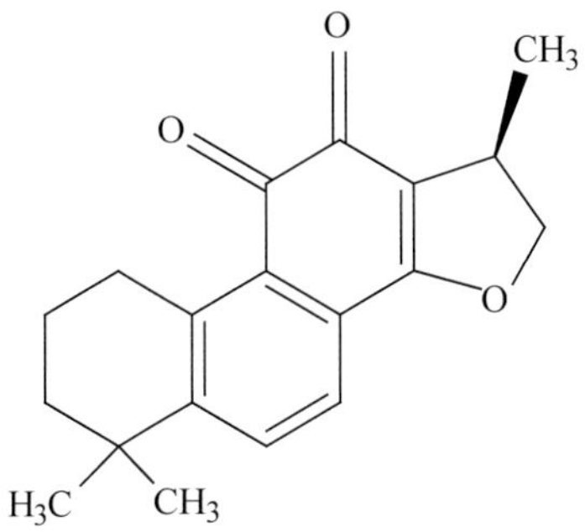

Fig. 54. Structure of crytotanshinone.

10.11 Podophyllotoxin

The synthesis of podophyllotoxin-loaded nanoparticles of the alkaloid has been achieved so as to ensure better bioavailability of the lignan in the treatment of genital warts (Chen et al. 2006).

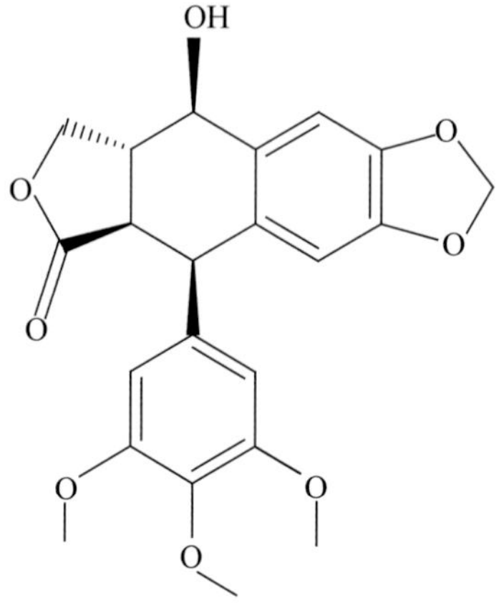

Fig. 55. Structure of podophyllotoxin.

10.12 Catechin

The free radical scavenging activity of catechin decreases dramatically when introduced in an alkaline medium. Chitosan encapsulated catechin particles have been developed to protect the compound (Zhang and Kosaraju 2006).

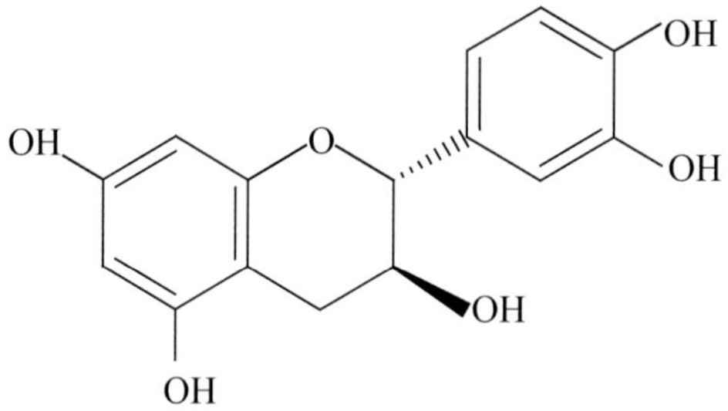

Fig. 56. Structure of catechin.

10.13 Triptolide

Poor solubility and toxicity have limiting effect on the therapeutic effects of the diterpene. In order to overcome the two obstacles, poly (DL-lactic acid) nanoparticles have been developed (Mei et al. 2005).

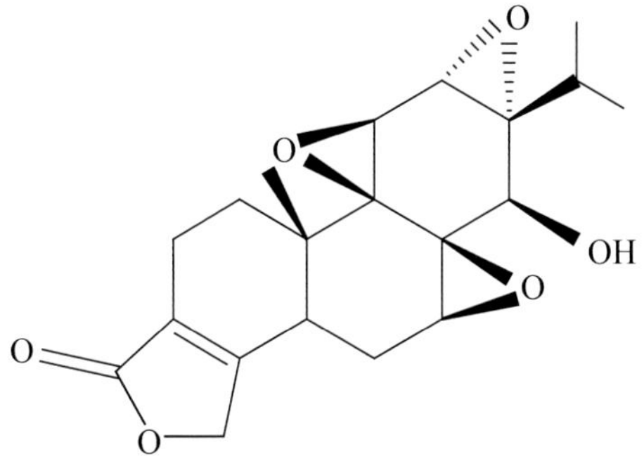

Fig. 57. Structure of triptolide.

10.14 Camptothecin

Glyco chitosan nanoparticles-encapsulated camptothecin (Fig. 58) have been synthesised for improved stability and anticancer action. The hydrophobic 5βcholanic acid moiety (Fig. 59) (hydrophobic) has been chemically conjugated with glycol chitosan (hydrophilic) backbone (Min et al. 2008).

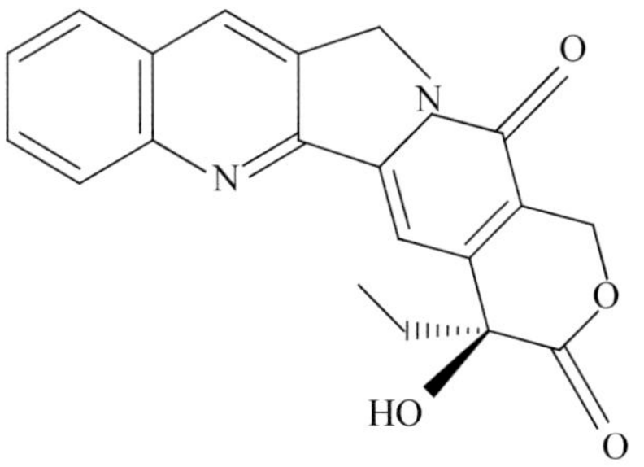

Fig. 58. Structure of camptothecin.

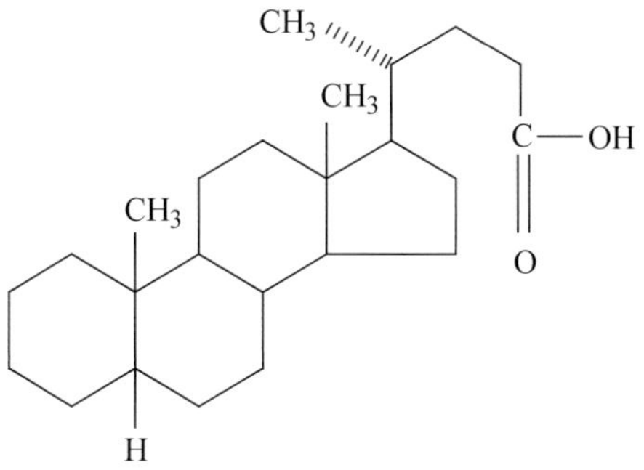

Fig. 59. Structure of 5βcholanic acid.

Further Reading

Al Shaal L, Shegokar R, Müller RH. Production and characterization of antioxidant apigenin nanocrystals as a novel UV skin protective formulation. *Int J Pharm*. 2011 25; **420:** 133–40.

Baskaran C. Green Synthesis of silver nanoparticles using Coleus forskohlii roots extract and its antimicrobial activity against Bacteria and Fungus. *Int J Drug Develop Res*. 2013; 5.

Chen H, Chang X, Du D, Liu W, Liu J, Weng T, Yang Y, Xu H, Yang X. Podophyllotoxin-loaded solid lipid nanoparticles for epidermal targeting. *J Control Release*. 2006; **110:** 296–306.

Gu Y, Wang Y, Zhou Q, Bowman L, Mao G, Zou B, Xu J, Liu Y, Liu K, Zhao J, Ding M. Inhibition of Nickel Nanoparticles-Induced Toxicity by Epigallocatechin-3-Gallate in JB6 Cells May Be through Down-Regulation of the MAPK Signaling Pathways. *PLoS One*. 2016 4; **11:** e0150954.

Hu LD, Xing Q, Meng J, Shang C. Preparation and enhanced bioavailability of cryptotanshinone-loaded solid lipid nanoparticles. *AAPS Pharm Sci Tech*. 2010; **11:** 582–587.

Li Y, Dong L, Jia A, Chang X, Xue H. Preparation and characterization of solid lipid nanoparticles loaded traditional Chinese medicine. *Int J Biol Macromolecules*. 2006; **38:** 296–299.

Mei Z, Li X, Wu Q, Hu S, Yang X. The research on the antiinflammatory activity and hepatotoxicity of triptolideloaded solid lipid nanoparticle. *Pharmacol Res*. 2005; **51:** 345–351.

Min KH, Park K, Kim YS, Bae SM, Lee S, Jo HG, Park RW, Kim IS, Jeong SY, Kim K, Kwon IC. Hydrophobically modified glycol chitosan nanoparticles-encapsulated camptothecin enhance the drug stability and tumor targeting in cancer therapy. *J Control Release*. 2008; **127:** 208–218.

Mitri K, Shegokar R, Gohla S, Anselmi C, Müller RH. Lutein nanocrystals as antioxidant formulation for oral and dermal delivery. *Int J Pharm*. 2011; 25; **420:** 141–6.

Sahu N, Soni D, Chandrashekhar B, Satpute DB, Saravanadevi S, Sarangi BK, Pandey RA. Synthesis of silver nanoparticles using flavonoids: hesperidin, naringin and diosmin, and their antibacterial effects and cytotoxicity. *Int Nano Lett*. 2016; **6:** 173–181.

Verma SK, Rastogil S, Arora I, Javed K, Akhtar M, Samim M. Nanoparticle Based Delivery of Quercetin for the Treatment of Carbon Tetrachloride Mediated Liver Cirrhosis in Rats. *J Biomed Nanotechnol*. 2016; **12:** 274–85.

Wang WP, Hul J, Sui H, Zhao YS, Feng J, Liu C. Glabridin nanosuspension for enhanced skin penetration: formulation optimization, *in vitro* and *in vivo* evaluation. *Pharmazie*. 2016; **71:** 252–7.

Yen FL, Wu TH, Lin LT, Cham TM, Lin CC. Naringenin-loaded nanoparticles improve the physicochemical properties and the hepatoprotective effects of naringenin in orally-administered rats with CCl(4)-induced acute liver failure. *Pharm Res*. 2009; **26:** 893–902.

Zhang L, Kosaraju SL. Biopolymeric delivery system for controlled release of polyphenolic antioxidants. *Eur Poly J*. 2007; **4:** 32956–32966.

CHAPTER 11

Nanoparticles of Herbal Extracts

Argemone mexicana L. (Papaveraceae)

Nanoparticles were characterized using ultraviolet–visible absorption spectroscopy, Fourier-transform infrared spectroscopy, X-ray diffraction and scanning electron microscope. The average particle size of the nanoparticles was found to be 30 nm. The nanoparticles showed high toxicity against different species of bacteria (Khandelwal et al. 2010).

Baliospermum montanum L. (Euphorbiaceae)

Water and ethanolic extracts of the plant were used for the preparation of the nanoparticles. The nanoparticles were evaluated for cytotoxic activity against prostate cancer and normal cell lines. Results of the *in vitro* investigations showed concentration and time dependent cytotoxic activity of nanoparticles against prostate cancer cells in 2 days (Cherian et al. 2015).

Cardiospermum halicacabum L. (Sapindaceae)

The reduction of the watery silver ions was carried out by exposing to *C. halicacabum* extract and the nanoparticles were prepared. Ultraviolet–visible absorption spectroscopy was used for qualitative characterization of the nanoparticles (Shekhawat et al. 2013).

Couroupita guianensis Aubl (Lecythidaceae)

The nanoparticles of the plant based on the flower extract showed antibacterial activity against *Pseudomonas aeroginosa* MTCC, *Bacillus subtilis* MTCC121, *Staphylococcus aureus* MTCC96, *Klebsiella pneumonia* MTCC109 and *E. coli* MTCC 912 (Sivakumar et al. 2015).

Cuscuta chinensis Lam. (Convolvulaceae)

A experimental study investigated the antioxidant and hepatoprotective activities of an ethanolic extract of the drug (dose 125 and 250 mg/kg, oral route) and nanoparticles (dose 25 and 50 mg/kg, oral route). The antioxidant and hepatoprotective activities of the nanoparticles was superior as compared to the ethanolic extract of the drug (Yen et al. 2008).

Drynaria fortunei J. Sm. Kunze (Polypodiaceae)

The 1000 µg/ml nanoparticles of water extract of the fern caused decrease in the cell viability. It also induced fragmentation of DNA and cell apoptosis. The profile of the 1000 µg/ml nanoparticles was superior as compared to the water extract (Hsu et al. 2011).

Erythrina indica Lam. (Fabaceae)

The silver nanoparticles were synthesised and demonstrated significant broad spectrum antibacterial activity against several bacteria. Further, the silver nanoparticles exhibited significant cytotoxic effect in breast and lung cancer cell lines (Rathi et al. 2015).

Euphorbia milii Des Moul (Euphorbiaceae)

The gold nanoparticles based on the methanolic extract of *E. milli* demonstrated significant muscle relaxant activity at doses of 10 mg/kg and 20 mg/kg. With the same doses of the gold nanoparticles, significant sedarive activity was also observed (Islam et al. 2015).

Houttuynia cordata Thunb (Saururaceae)

A study was undertaken with respect to enhancement of dermal penetration of water soluble extract of the fish smell herb. The dermal penetration may enhance the therapeutic efficacy against atopic dermatitis. The extract has suppressive action on the production of immunoglobulin E. It promoted the expression of interferon gamma (Kwon and Kim 2014).

Kaempferia parviflora Wall. ex Baker (Zingiberacerae)

A skin model (human derived epidermal keratinocytes) was used for comparative permeability study of topically applied Thai Ginseng loaded solid nanoparticles and a gel formulation based on the extract. A marked difference was observed between the nanoparticles and the gel with regard to permeation of the total flavonoids. The permeation was definitely better in case of the nanoparticles (Sutthanut et al. 2009).

Libanotis buchtomensis

The developed nanoparticles were successfully utilised for selective separation of osthole from the extract of *L. buchtomensis* (He et al. 2016).

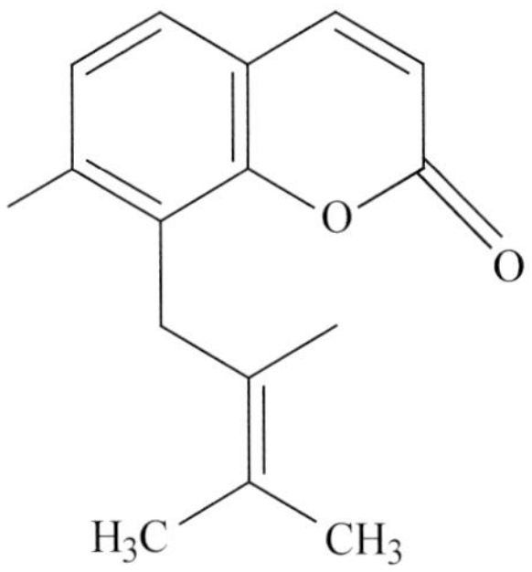

Fig. 60. Structure of osthole.

Lonicera japonica Thunb (Caprifoliaceae)

The antimicrobial activity of the silver nanoparticles-*L. japonica* was evaluated against *Escherichia coli* CMCC44113. The combination yielded better antimicrobial activity as compared to either silver nanoparticles or extract of honeysuckle (Yang et al. 2016).

Morinda coreia Buck. Ham (Rubiaceae)

Silver nanomaterials developed from leaf and fruit water extracts of *M. coreia* demonstrated significant antimicrobial activity against five human pathogenic bacteria (Kannan et al. 2014).

Passiflora serratodigitata L. (Passifloraceae)

The crude extract of the leaves and stem, ethylacetate fraction, and residual water fraction demonstrated promising gastroprotective activity. The nanocapsule loaded with ethylacetate fraction showed 10-fold more gastroprotective activity than the ethylacetate fraction (Strasser et al. 2014).

Picrorhiza kurroa Royle ex Benth. (Plantaginaceae)

The herb is a well known hepatoprotective herb used in Ayurveda. In order to improve bioavailability of picrosides I (Fig. 61) and II (Fig. 62), a nanoformulation based on *P. kurroa* extract was developed by nanoprecipitation method. The method showed high encapsulation efficiency as $60.1 \pm 2.8\%$ for picroside I and $67.2 \pm 7.4\%$ for picroside II (Jiaa et al. 2015).

Picroliv is a mixture of picroside I and kutkoside (Fig. 63). The picroliv-loaded poly lactic acid nanoparticles demonstrated more cytotoxic activity on KB cells as compared to the crude drug (Guliani et al. 2016).

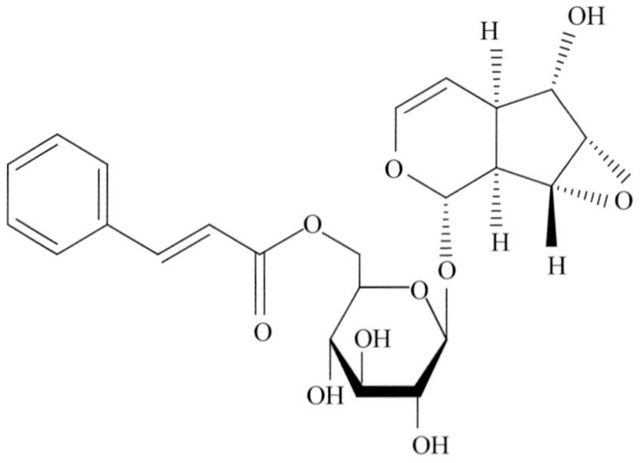

Fig. 61. Structure of picroside I.

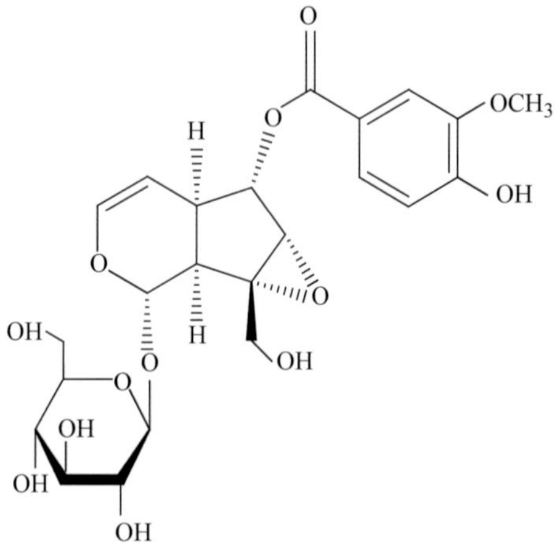

Fig. 62. Structure of picroside II.

Plectranthus amboinicus (Lour.) Spreng (Lamiaceae)

The synthesized silver nanoparticles exhibited better antimicrobial activity against *Escherichia coli* and *Penicillium* spp. (Ajitha et al. 2014).

Sesbania grandiflora L. (Fabaceae)

The silver nanoparticles exhibited promising antibacterial activity against *Salmonella enterica* and *Staphylococcus aureus* (Das et al. 2013).

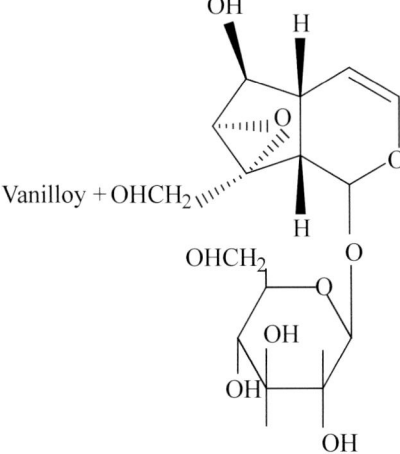

Fig. 63. Structure of kutkoside.

Tephrosia purpurea L. (Fabaceae)

The sliver nanoparticles using the leaf extract of *T. purpurea* showed selective antibacterial activity against *Penicillium* spp. and *Pseudomnas* spp. (Ajitha et al. 2014).

Turnera ulmifolia L. (Passifloraceae)

The nanoparticles showed fair antibacterial activity against several disease causing microorganisms (Shekhawat et al. 2012).

Further Reading

Ajitha B, Ashok Kumar Reddy Y, Sreedhara Reddy P. Biosynthesis of silver nanoparticles using *Plectranthus amboinicus* leaf extract and its antimicrobial activity. *Spectrochim Acta A Mol Biomol Spectrosc.* 2014a; **128:** 257–62.

Ajitha B, Reddy YA, Reddy PS. Biogenic nano-scale silver particles by *Tephrosia purpurea* leaf extract and their inborn antimicrobial activity. *Spectrochim Acta A Mol Biomol Spectrosc.* 2014b; **121:** 164–72.

Cherian AM, Snima KS, Kamath CR, Nair SV, Lakshmanan VK. Effect of *Baliospermum montanum* nanomedicine apoptosis induction and anti-migration of prostate cancer cells. *Biomed Pharmacother.* 2015; **71:** 201–9.

Das J, Paul Das M, Velusamy P. *Sesbania grandiflora* leaf extract mediated green synthesis of antibacterial silver nanoparticles against selected human pathogens. *Spectrochim Acta A Mol Biomol Spectrosc.* 2013; **104:** 265–70.

Guliani A, Kumari A, Kumar D, Yadav SK. Development of nanoformulation of picroliv isolated from *Picrorrhiza kurroa. IET Nanobiotechnol.* 2016; **10:** 114–9.

He G, Tang Y, Hao Y, Shi J, Gao R. Preparation and application of magnetic molecularly imprinted nanoparticles for the selective extraction of osthole in *Libanotis buchtomensis* herbal extract. *J Sep Sci.* 2016; **39:** 2313–20.

Hsu CK, Liao MH, Tai YT, Liu SH, Ou KL, Fang HW, Lee IJ, Chen RM. Nanoparticles prepared from the water extract of Gusuibu (*Drynaria fortunei* J. Sm.) protects osteoblasts against insults and promotes cell maturation. *Int J Nanomedicine.* 2011; **6:** 1405–13.

Islam NU, Khan I, Rauf A, Muhammad N, Shahid M, Shah MR. Antinociceptive, muscle relaxant and sedative activities of gold nanoparticles generated by methanolic extract of *Euphorbia milii*. *BMC Complement Altern Med*. 2015; **15:** 160.

Jiaa D, Barwalb I, Thakurb S, Yadav SC. Methodology to nanoencapsulate hepatoprotective components from *Picrorhiza kurroa* as food supplement. *Food Bioscience.* 2015; **9:** 28–35.

Kannan N, Shekhawat MS, Ravindran CP, Manokari M. Preparation of silver nanoparticles using leaf and fruit extracts of *Morinda coreia* Buck., Ham. –A green approach. *J Sci Innov Res* 2014; **3:** 315–318.

Khandelwal N, Singh A, Jain D, Upadhyay MK, Verma HN. Green synthesis of silver nanoparticles using *Argemone mexicana* leaf extract and evaluation of their antimicrobial activities. *J Nanomater Biostruct*. 2010; **5:** 483–489.

Kwon TK, Kim JC. *In vitro* skin permeation and anti-atopic efficacy of lipid nano carriers containing water soluble extracts of *Houttuynia cordata*. *Drug Dev Ind Pharm*. 2014; **40:** 1350–7.

Rathi Sre PR, Reka M, Poovazhagi R, Arul Kumar M, Murugesan K. Antibacterial and cytotoxic effect of biologically synthesized silver nanoparticles using aqueous root extract of *Erythrina indica* Lam. *Spectrochim Acta A Mol Biomol Spectrosc*. 2015; **135:** 1137–44.

Shekhawat MS, Kannan N, Manokari M. Biogenesis of silver nanoparticles using leaf extract of *Turnera ulmifolia* Linn. and screening of their antimicrobial activity. *J Ecobiotechnol*. 2012; **4:** 54–57.

Shekhawat MS, Manokari M, Kannan N, Revathi J, Latha R. Synthesis of silver nanoparticles using *Cardiospermum halicacabum* L. leaf extract and their characterization. *J Phytopharmacol.* 2013; **2:** 15–20.

Sivakumar T, Rathimeena TS, Ponmanickam P. Production of silver nanoparticles synthesis of *Couroupita guianensis* plant extract against human pathogen and evaluations of antioxidant properties. *Int of Life Sci.* 2015; **3:** 333–340.

Strasser M, Noriega P, Löbenberg R, Bou-Chacra N, Bacchi EM. Antiulcerogenic potential activity of free and nanoencapsulated *Passiflora serratodigitata* L. extracts. *Biomed Res Int*. 2014; **2014:** 434067.

Sutthanut K, Lu X, Jay M, Sripanidkulchai B. Solid lipid nanoparticles for topical administration of *Kaempferia parviflora* extracts. *J Biomed Nanotechnol*. 2009; **5:** 224–32.

Yang L, Aguilar ZP, Qu F, Xu H, Xu H, Wei H. Enhanced antimicrobial activity of silver nanoparticles-*Lonicera Japonica* Thunb combo. *IET Nanobiotechnol*. 2016; **10:** 28–32.

Yen FL, Wu TH, Lin LT, Cham TM, Lin CC. Nanoparticles formulation of *Cuscuta chinensis* prevents acetaminophen-induced hepatotoxicity in rats. *Food Chem Toxicol*. 2008; **46:** 1771–7.

PART B

PHYTOCANNABINOIDS

CHAPTER 12

Cannabis indica—Historical Aspects

12.1 Introduction

Cannabis as a medicine was used before the Christian era in Asia, mainly in India. The introduction of cannabis in the Western medicine occurred in the midst of the 19th century, reaching the climax in the last decade of that century, with the availability and usage of cannabis extracts or tinctures (Lambert 2001). In the first decades of the 20th century, the Western medical use of cannabis significantly decreased largely due to difficulties to obtain consistent results from batches of plant material of different potencies. The identification of the chemical structure of cannabis components and the possibility of obtaining its pure constituents were related to a significant increase in scientific interest in such a plant, since 1965. This interest was renewed in the 1990s with the description of cannabinoid receptors and the identification of an endogenous cannabinoid system in the brain (Zuardi 2006).

12.2 Cannabis in Traditional Chinese Medicine

It is estimated that around 6000 BCE seeds of Cannabis were used as in China (Joe et al. 1993). The first recorded use for medicinal purpose in China pharmacopoeia dates back to 2727 BCE. It was in 1500 BCE that Cannabis was cultivated in China. The *pen-ts'ao ching,* world's oldest pharmacopoeia has described the use of Cannabis in the treatment of rheumatism, malaria, diseases of the female genital tract and constipation (Hou 1977). With regard to the, treatment of constipation, a study reported the efficacy of a Chinese herbal proprietary medicine (Hemp Seed Pill) in the treatment of functional constipation (Whiting and Ford 2011). Hua T'o, the father of Chinese surgery, used a compound derived from Cannabis alongwith wine for inducing anaesthesia in patients undergoing surgical procedures (Li and Lin 1974; Auvinen and Peltola 1999).

12.3 Cannabis in Indian Medicine

The Atharva Veda written sometime around 1200–800 BCE, has mentioned Cannabis to be one of the five sacred plants. Here Cannabis refers to Bhang (dried leaves, stem and seeds). *Anandakanda,* the ancient Indian alchemy text, has described 10 types of Cannabis users. Further, 50 other different preparations of Cannabis have also been described for the purpose of rejuvenation, aphrodisiac effect and cure of several diseases. *Ananda kanda* has given a detailed description of the toxic effects that appear in human beings in nine successive stages (Ethan 2005).

In Ayurvedic works, about 51 important formulations of different categories containing Cannabis have been described (Dominik et al. 2003). The Cannabis root has been described to be a poison in *Susruta Samhita* (700 B.C.) and in many medieval Ayurvedic works on Materia Medica it has been described as an upavisa (poison of minor importance). Sushruta has indicated the use of Cannabis during surgery (Wujastyk 2002).

Sharangadhara Samhita, a compendium of therapeutics (thirteenth century A.D.), has included medicaments titrated with fresh extract of *bhang.* Authoritative *Ayurvedic* works on Materia Medica such as *Dhanwantari nighantu* (eighth century A.D.), *Madanapala nighantu* (1374 A.D.) and *Rajanighantu* (1450 A.D.) have described the properties, actions and indications of both Cannabis and opium.

Bhavamishra (fifteenth century A.D.), a contemporary of Paracelsus, in his compendium on medicine and therapeutics, *Bhavaprakasha,* has described the properties, actions, indications and formulations of Cannabis (Dwarakanath 1965; Chaturvedi et al. 1981). *Bhavaprakasha* described Cannabis as antiphlegmatic, digestive, bile affecting, pungent, and astringent, prescribing it to stimulate the appetite, improve digestion, and better the voice.

12.4 Cannabis in Traditional Arabic and Persian Medicine

Arab physicians knew and used its, anti-emetic, anti-epileptic, anti-inflammatory, analgesic antipyretic and diuretic properties (Lozano 1997). Dioscorides (first century A.D.), who had described both Cannabis and opium, is seen to have made use of them for therapeutic purposes. Galien (138–201 A.D.) and Rhazes (865–925 A.D.) have given detailed descriptions of these drugs, their actions, therapeutics and uses.

Zoroaster wrote a multi-volume ancient Persian religious text known as The Zoroastrian Zend-Avesta between 559 BCE and 379 CE. In this work, Zoroaster has mentioned Bhanga as a valuable narcotic. In addition, Bhanga has also been described as an abortifacient. The credit of introducing Cannabis in Europe goes to Scythian tribes around 500 BCE. As per estimates, it was between 500 BCE–1000 BCE that Cannabis had spread throughout northern Europe.

Authoritative Arabic and Persian medical works such as (1) *Firdo usul-Hikmat* and (2) *Mujardat Quanan* have not only described the properties of these drugs, but have also included a number of formulations containing them. It would appear that potions containing Cannabis and linctus containing opium were popular in Arabia, Persia and Muslim India (Iorno 2001).

12.5 Cannabis in Herbal Materia Medica and Western Pharmacopoeia

Dioscorides (77 CE) described properties of Cannabis in *'De Materia Medica'*. In his third book, Dioscorides describes that the root of Cannabis when boiled and applied reduces pain, oedema and inflammation in inflamed joints. The seeds consumed in certain quantity supress conception. The expressed juice can be used in the treatment of earaches (Osbaldeston 2000).

Although there is evidence that Cannabis was used in Europe from the thirteenth century, after Marco Polo returned from his journey to the East in 1297, its medical use became more popular in the nineteenth century, when the British physician William B. O'Shaugnessy brought back an account of the remarkable effects of this plant from India. Even Queen Victoria is said to have sipped marijuana tea prescribed by her court physician to treat menstrual cramps (Frazzetto et al. 2003).

Cannabis has been used for centuries for both symptomatic and prophylactic treatment of migraine. It was highly esteemed as a headache remedy by the most prominent physicians of the ages between 1874 and 1942. It lasted as a remedy for migraine in Western pharmacopoeia till the mid-twentieth century (Russo 1998).

Squire in *Companion to the Latest Edition of the British Pharmacopoeia* has described Cannabis as sedative, anodyne, and hypnotic (Squire 1899). In *Merck's 1899 Manual of the Materia Medica*, a preparation made with Cannabis extracts named *Cannabine Tannate Merck* is recommended as a hypno-sedative in the treatment of hysteria, delirium, nervous insomnia (Rahway 1930).

Cushny (1906) has defined Cannabis as a hypnotic agent in *A Textbook of Pharmacology and Therapeutics.* The 1907 and 1930 editions of *Merck Indexes* have defined that Cannabis is a "hypnotic property of Cannabis (Rahway 1930). *Bruce and Dilling's Materia Medica and Therapeutics* also maintains the hypnotic property of Cannabis (Dilling 1933). *The British Pharmaceutical Codex,* 1934 has again mentioned Cannabis as "sedative or hypnotic" (Pharmaceutical Society of Great Britain 1934).

Cannabis is listed as hypnotic and sedative in the 1940 edition of the *Merck Manual of Therapeutics and Materia Medica* ("Rahway" 1940). The pharmaceutical trade journal *Ciba Symposia* in 1946 (Robinson 1946), also points out that Cannabis "is sometimes employed as a hypnotic drug in those cases where opium, because of long-continued use, has lost its efficiency". *MerckIndex* (O'Neil et al. 2001), continued to list drowsiness as an effect of Cannabis smoking/inhalation.

12.6 Government Reports

12.6.1 Indian Hemp Drugs Commission Report

Indian Hemp Drugs Commission Report was initiated by the British in the year of 1894. It noted sleep to be the major effect of Cannabis (Kaplan 1969; Kalant 1972). In some people, Cannabis was reported to produce peculiar loss of sensation of space and time. This is followed by the stage of deep sleep. The intoxication lasts for about three hours, when sleep supervenes. The Commission also noted slight narcotic effects more or less complete (Mikuriya 1968).

12.6.2 La Guardia Committee Report

Fifty years later, the *La Guardia Committee Report*, after an in-depth investigation of smoked Cannabis in the United States reported prolonged drowsiness as a major effect (Mayor's Committee on Marihuana 1944).

12.6.3 The Wootton Report

In order to investigate the effects of Cannabis, the British in the year of 1968 published the *Wootton Report*. The report emphasised its calming and relaxing effects (U.K. Home Office 1968).

12.6.4 The Le Dain Commission

In order to investigate the effects of Cannabis, Canada in the year of 1972 appointed the *Le Dain Commission*. The *Commission* published the calming effect of Cannabis (Le Dain 1970).

12.6.5 The Shafer Commission

Nearly simultaneously, President Richard Nixon commissioned a report to study Cannabis abuse in the U.S., commonly referred to as the *Shafer Commission*. The *Shafer Commission like The Wootton Report* and *The Le Dain Commission,* reported relaxation as a major effect of Cannabis (Shafer 1972).

12.6.6 Marihuana Tax Act of 1937 (MTA)

The Marihuana Tax Act of 1937 was a US Act that placed tax on the sale of Cannabis. The act was drafted by Harry Jacob Anslinger (the first commissioner of the U.S. Treasury Department's Federal Bureau of Narcotics (FBN)).

12.7 Cannabis and The American Herbal Pharmacopoeia

The American Herbal Pharmacopoeia has recently published two-part "Cannabis monograph". The monograph has been written and reviewed by the world's leading experts. It tends to bring together scientific data and issues long-awaited standards for the plant's identity, purity, quality, and botanical properties. The monograph gives doctors who want to prescribe cannabis therapy a full scientific understanding of the plant, its constituent components, and its biologic effects.

Further Reading

Auvinen A, Peltola J. On oenodotes and oenotherapy-wine in medicine. *Duodecim*. 1999; **115:** 2623–32.
Chaturvedi GN, Tiwari SK, Rai NP. Medicinal use of opium and cannabis in medieval India. *Indian J Hist Sci*. 1981; **16:** 31–5.
Cushny A. Cannabis indica. In *A textbook of pharmacology and therapeutics or the actions of drugs in health and disease* (1906; 4th ed., pp. 232–234). New York: Lea Brothers & Co.

Dilling WJ. Cannabis indica. In *Bruce and Dilling's Materia Medica and Therapeutics an Introduction to the Rational Treatment of Disease* (1933; 14th Rev. ed., p. 383). London: Cassell and Company, Limited.

Dominik W, Gupta RD, Reynolds L, Brady CM. History of medicinal use of Cannabis in ancient India. *J Urol.* 2003; **169:** 253.

Dwarakanath C. Use of opium and Cannabis in the traditional systems of medicine in India. *Bull Narcotics.* 1965; 17, 1, January–March, W.H.O., Geneva.

Ethan R. Cannabis in India: ancient lore and modern medicine. In R. Mechoulam (Ed.), 2005; 1–22.

Frazzetto G. Does marijuana have a future in pharmacopoeia? *EMBO Rep.* 2003; **4**(7): 651–653.

Hou JP. The development of Chinese herbal medicine and the Pen-ts'ao. *Comp Med East West.* 1977; **5:** 117–22.

Iorno I. The therapeutic uses of *Cannabis sativa* (L.) in Arabic medicine. *J Cannabis Ther.* 2001; **1:** 63–70.

Joe Z, Stark H, Seligman J, Levy R, Werker E, Breuer A, Mechoulam R. Early medical use of Cannabis. *Nature.* 1993; **363:** 215.

Kalant OJ. Report on the Indian hemp drugs commission 1893–94; a critical review. *Int J Addict.* 1972; **7:** 177–96.

Kaplan J. Marijuana-Report of the Indian Hemp Drugs Commission, 1893–1894. Thomas Jefferson Publishing Co., Silver Spring, MD, 1969.

Lambert DM. Medical use of cannabis through history. *J Pharm Belg.* 2001; **56:** 111–8.

Le Dain G. *Interim report of the Commission of Inquiry into the non-medical use of drugs.* Ottawa, ON: Information Canada, 1970.

Li HL, Lin H. An archaeological and historical account of cannabis in China. *Econ Bot.* 1974; **28:** 437–47.

Lozano I. Therapeutic use of *Cannibis sativa* L. in Arab medicine. *Asclepio.* 1997; **49:** 199–208.

Mayor's Committee on Marihuana. (1944). *The marihuana problem in the city of New York: Sociological, medical, psychological and pharmacological studies.* Lancaster, PA: The Jaques Cattell Press.

Mikuriya TH. Physical, mental, and moral effects of marijuana: The Indian Hemp Drugs Commission Report. *Int J Addict.* 1968; **3:** 2.

O'Neil, M J. et al. (Eds.). Cannabis. In *The Merck Index. An Encyclopedia of Chemicals, Drugs, and Biologicals* (2001; 13th ed., p. 292). Whitehouse Station, NJ: Merck & Co., Inc.

Osbaldeston TA. De Materia Medica: Being an Herbal with many other medicinal materials, 2000 (Publisher Ibidis Press: Johannesburg).

Pharmaceutical Society of Great Britain. Cannabis. In *The British Pharmaceutical Codex, 1934:An Imperial Dispensatory for the Use of Medical Practitioners and Pharmacists* (1934; p. 270). London: The Pharmaceutical Press.

Rahway NJ. Cannabis. In *Merck's Index. An Encyclopedia for the Chemist, Pharmacist and Physician.* 1930 (4th ed.,p. 147). Merck & Co., Inc.

Rahway NJ. Cannabis. In *The Merck Manual of Therapeutics and Material Medica. A Source of Ready Reference for the Physician* (1940, 7th ed., p. 1356). Merck & Co., Inc.

Robinson R. *The great book of Hemp: The complete guide to the environmental, commercial, and medicinal uses of the world's most extraordinary plant.* Rochester, VT: Park Street Press, 1996.

Russo E. Cannabis for migraine treatment: the once and future prescription? An historical and scientific review. *Pain.* 1998; **76:** 3–8.

Shafer RP. *Marihuana—A signal of misunderstanding. The official report of the National Commission on Marihuana and Drug Abuse.* New York, NY: Signet, 1972.

Squire PW. *Cannabis indica.* Indian hemp. In *Companion to the Latest Edition of the British Pharmacopoeia* (1899; 17th ed., pp. 179–181). London: J. & A. Churchill.

U.K. Home Office. (1968). *Cannabis: Report by the advisory committee on drug dependence.* Home Office, Her Majesty's Stationery Office.

Whiting RL, Ford AC. Efficacy of traditional chinese medicine in functional constipation. *Am J Gastroenterol.* 2011; **106:** 1003–4.

Wujastyk D. Cannabis in traditional Indian herbal medicine. In A. Salema (Ed.), 2002; 45–73.

Zuardi AW. History of cannabis as a medicine: a review. *Rev Bras Psiquiatr.* 2006; **28:** 153–7.

Chapter 13

Botany of Cannabis

13.1 *Cannabis sativa* L.

13.1.1 Distribution

C. sativa is widespread throughout North America, grows in the wilderness of Northern India and is also found in Southern Siberia China.

13.1.2 Botany

C. sativa is a herb, the erect stems growing from 3 to 10 feet or higher, very slightly branched, having greyish-green hair. The leaves are palmate, with five to seven leaflets (three on the upper leaves), numerous, on long thin petioles with acute stipules at the base, linear-lanceolate, tapering at both ends, the margins sharply serrate, smooth and dark green on the upper surface, lighter and downy on the under one. The small flowers are unisexual, the male having five almost separate, downy, pale yellowish segments, and the female a single, hairy, glandular, five-veined leaf enclosing the ovary in a sheath. The ovary is smooth, one-celled, with one hanging ovule and two long, hairy thread-like stigmas extending beyond the flower for more than its own length. The fruit is small, smooth, light brownish-grey in colour, and completely filled with the seed.

13.2 *Cannabis indica* Lam

13.2.1 Distribution

India, Afghanistan, Bangladesh and Pakistan.

13.2.2 Botany

The typical example of *C. Indica* is a more compact, thick-stemmed bush than its cousins, usually reaching a height of less than two metres. The foliage is generally a dark shade of green. Some of them also appear to have almost blue or green-black leaves. These leaves are composed of short, wide blades. Indica strains tend to produce more side-branches and denser overall growth than Sativas, resulting in wider, bushier plants. Indica flowers form in thick clusters around the nodes of the female plant (the points at which pairs of leaves grow from the stem and branches). They usually weigh more than Sativa flowers of similar size, as they are more solid 2.3: *C. sativa* ssp. *ruderalis.*

13.3 *Cannabis ruderalis* Janisch. (Wild hemp)

13.3.1 Distribution

Central Russia.

13.3.2 Botany

C. ruderalis grows to a height of only 60 cm. It has few branches and small leaves. The inflorescences are small and form on the end of the stalk.

Chapter 14

Chemistry of *Cannabis indica*

14.1 Resin

Cannabin stands for a biologically active resin extracted from Indian hemp. Smith (1846) reported about the presence of Cannabin in *Cannabis indica*. The resin is composed of cannabinol, pseudo-cannabinol, cannabinin (Mechoulam 1986). Now-a-days, the resin is considered to be residual THC and plant matter. The *cannabin* is the alcoholic or resinous extract employed in medicine (Joyce and Curry 1970; Mechoulam and Hanuš 2000).

14.2 The Volatile Oil

Personne (1857) separated the volatile oil in cannabene and *cannabene hydride.* Cannabene is a colourless fluid having intoxicating properties similar to *Cannabis.*

14.3 Alkaloids

14.3.1 Cannabinine

Cannabinine was isolated by Siebold and Bradbury (1881), in a very small quantity. The extraction methodology was similar to that of nicotine.

14.3.2 Tatano-cannabinine

Matthew Hay (1883) isolated tatano-cannabinine in a crystalline form having tetanic property. Jahns in 1889 proposed tatano-cannabinine identical with choline.

14.3.3 Cannabamines A-D

Their structure is not yet confirmed (El Feraly and Turner 1975).
   Choline, trigonelline, muscarine, and an unidentified betain: these have been reported alongwith cannabamines.

14.3.4 Cannabisativine

An ethanol extract of the root of a Mexican variant of *C. sativa* afforded, after partitioning and chromatography, the new spermidine alkaloid cannabisativine (Turner et al. 1976).

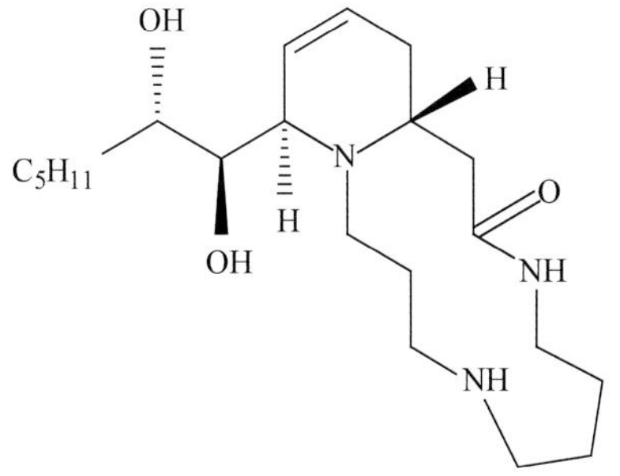

Fig. 64. Structure of cannabisativine.

14.3.5 Anhydrocannabisativine

Ethanol extracts of the leaves and roots of a Mexican variant of *C. sativa* afforded the new spermidine alkaloid, anhydrocannabisativine (Elsohly et al. 1978).

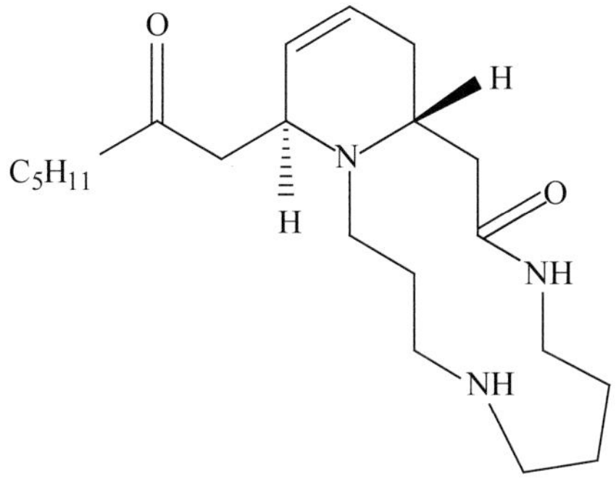

Fig. 65. Structure of anhydrocannabisativine.

14.4 Cannabinoids

14.4.1 Introduction

Cannabinoids are diverse chemical compounds that were first discovered in 1964. They were originally defined as a group of C21 compounds uniquely produced by the cannabis plant. 113 cannabinoids have been isolated from flower, leaf and stem of

Cannabis plant. Since they exist naturally in the cannabis plant, they are also known as phytocannabinoids. They have been addressed as natural cannabinoids, herbal cannabinoids and classical cannabinoids. Tetrahydrocannabinol is the most important cannabinoid (Argurell et al. 1984; Appendino et al. 2011). Cannabichromene is found in other plants also.

14.4.2 Classification

Cannabinoids, a class of meroterpenoids derived from the alkylation of an olivetol-like alkyl resorcinol with a monoterpene unit, are the most typical constituents of *Cannabis*. This class includes over a hundred members belonging to several structural types, mainly differing by the constitution of their terpenoid moiety (Makriyannis and Rapaka 1987; Appendino and Taglialatela-Scafat 2013). Each of the 113 cannabinoid compounds falls within one of six different subclasses. These six cannabinoid subclasses include:

- Tetrahydrocannabinols (THCs)
- Cannabierols (CBGs)
- Cannabinodiols and cannabinols (CBDL and CBNs)
- Cannabidiols (CBDs)
- Cannabichromenes (CBCs)
- Miscellaneous cannabinoids – cannabitriol (CBT), cannabicyclol (CBL), cannabielsoin (CBE) and others.

Another way of classification is as follows:

- Phytocannabinoids occur uniquely in the cannabis plant.
- Endogenous cannabinoids are produced in the body of humans and other animals.
- Synthetic cannabinoids are similar compounds produced in a laboratory.

14.4.3 Important cannabinoids

14.4.3.1 Tetrahydrocannabinol (THC)

Tetrahydrocannabinol is more precisely known as $(-)$-*trans*-Δ^9-tetrahydrocannabinol. It is the chief psychoactive constituent of cannabis and hasish. Tetrahydrocannabinol is found in the resin secreted by glands of the cannabis plant. The buds are often

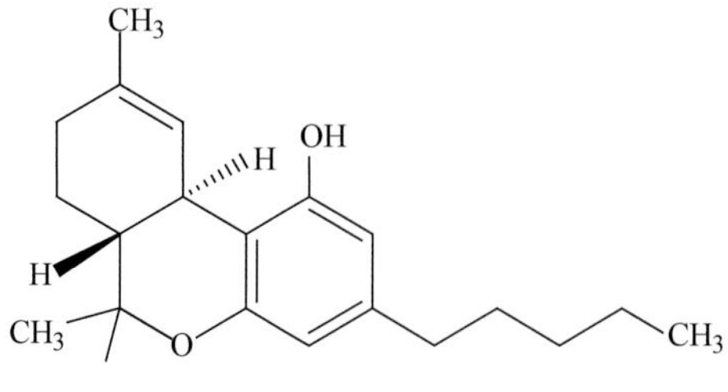

Fig. 66. Structure of tetrahydrocannabinol.

preferred because of their higher tetrahydrocannabinol content. The credit of isolation of tetrahydrocannabinol goes to Raphael Mechoulam, Isreal-based Scientist.

The role of THC in Cannabis, seems, to protect the plant from herbivores or pathogens. Initially, tetrahydrocannabinol was included in the schedule I of the 1971 Convention of Psychotropic Substances. Recently, World Health Organisation (WHO) reclassified tetrahydrocannabinol in the less stringent schedule III.

14.4.3.2 Cannabidiol (CBD)

Cannabidiol accounts for 40% of the cannabis extract.

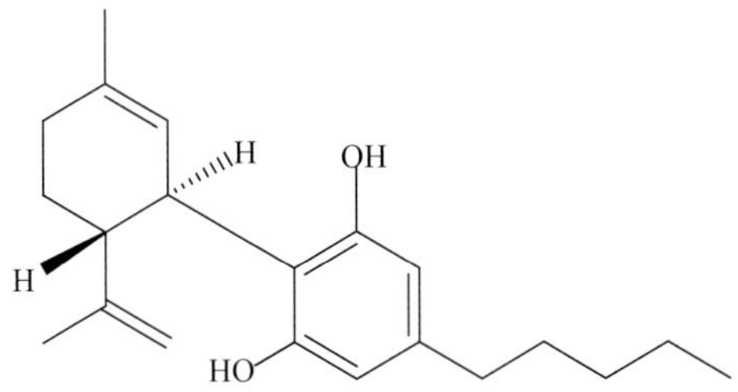

Fig. 67. Structure of cannabidiol.

14.4.3.3 Cannabinol (CBN)

Cannabinol is degradation product of tetrahydrocannabinol.

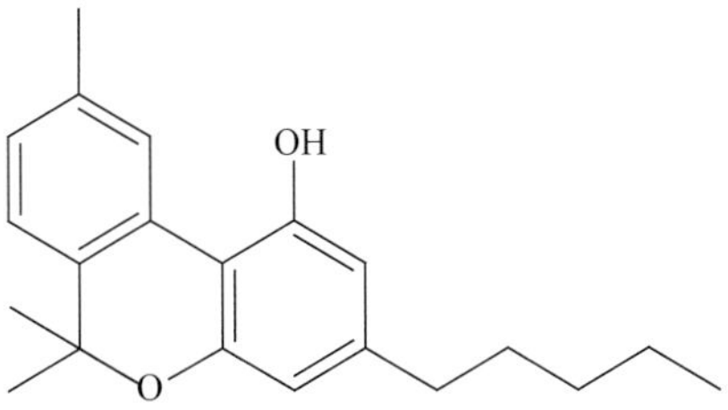

Fig. 68. Structure of cannabinol.

14.4.3.4 Cannabichromene (CBC)

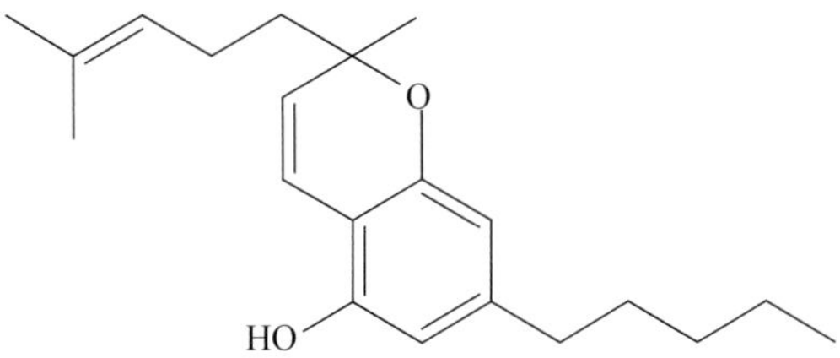

Fig. 69. Structure of cannabichromene.

14.4.3.5 Cannabigerol (CBG)

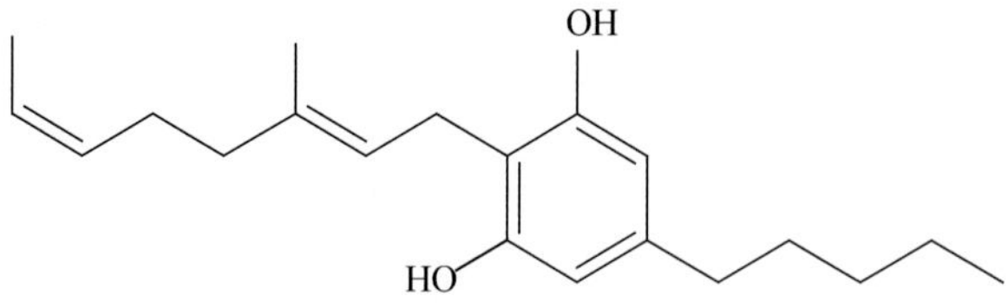

Fig. 70. Structure of cannabigerol.

14.4.3.6 Tetrahydrocannabitriol

14.4.3.7 Cannabidiolic acid (CBDA)

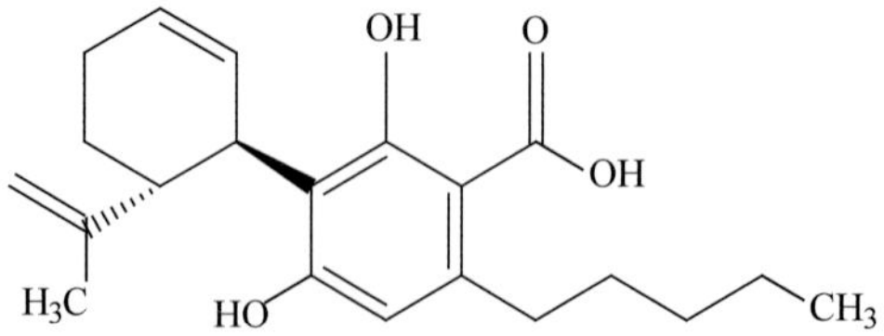

Fig. 71. Structure of cannabidiolic acid.

14.4.3.8 Tetrahydrocannabivarin (THCV)

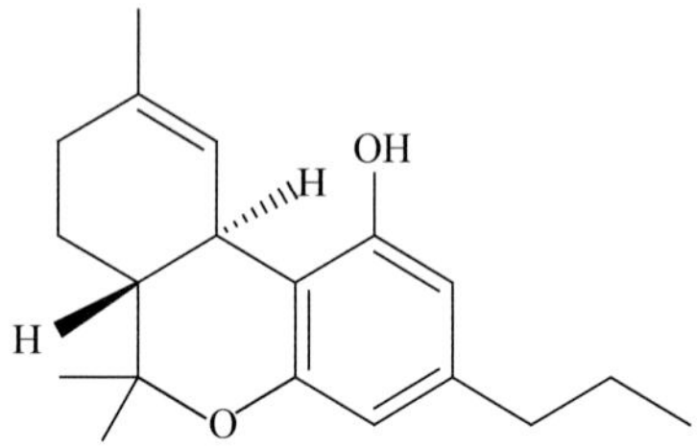

Fig. 72. Structure of tetrahydrocannabivarin.

14.4.3.9 Cannabivarin (CV)

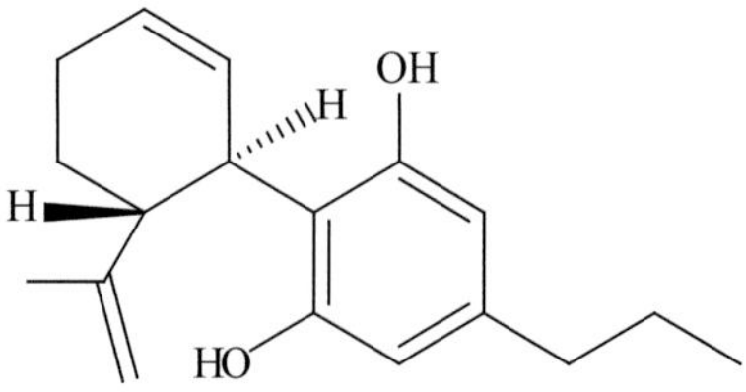

Fig. 73. Structure of cannabivarin.

Further Reading

Appendino G, Chianese G, Taglialatela-Scafati O. Cannabinoids: occurrence and medicinal chemistry. *Curr Med Chem*. 2011; **18:** 1085–99.

Appendino G, Taglialatela-Scafat O. Cannabinoids: Chemistry and Medicine. *Nat Prod*. 2013; 3415–3435.

Argurell S, Dewey WL, Wilette RE (Eds.). The Cannabinoids: Chemical, Pharmacologic and Therapeutic Aspects. Academic Press, Orlando, 1984.

El Feraly FS, Turner CE. Alkaloids of *Cannabis sativa* leaves. *Phytochemistry*. 1975; **14:** 2304.

Elsohly MA, Turner CE, Phoebe CH, Knapp JE, Schiff PL, Slatkin DJ. Anhydrocannabisativine, a new alkaloid from *Cannabis sativa* L. *J Pharm Sci*. 1974; **67**(1): 124.

Joyce CBR, Curry SH. The botany and chemistry of Cannabis. Churchill, London, 1970 .

Makriyannis A, Rapaka RS. The medicinal chemistry of cannabinoids: an overview. *NIDA Res Monogr*. 1987; **79:** 204–10.

Mechoulam R, Hanuš L. A historical overview of chemical research on cannabinoids. *Chem Phys Lipids*. 2000; **108:** 1–13.

Mechoulam R. The pharmacohistory of *Cannabis sativa*. *In*: R. Mechoulam (Ed.), 1986; 1–19 .

Turner CE, Hsu MH, Knapp JE, Schiff PL, Slatkin DJ. Isolation of cannabisativine, an alkaloid, from *Cannabis sativa* L. root. *J Pharm Sci*. 1976; **65**(7): 1084–5.

Chapter 15

General and Clinical Pharmacology of *Cannabis* sp.

15.1 Earlier Reports

Observations on the medicinal properties of the *C. sativa* growing in India were reported in 1843 (Clendinning 1843). The use of *C. indica* in the treatment of tetanus, hydrophobia and cholera was reported (Shaw 1843; Inglis 1845). A case reporting efficacy of the tincture of *C. indica* in the treatment of dysmenorrhoea was reported (Barrow 1847). Possible treatment of certain type of headache with Indian hemp have been described (Mackenzie 1887).

Marihuana activity of cannabinol was documented in 1945 (Lowew 1945). Physiologically active fraction of *C. sativa* was described (Bose and Mukerji 1945). A study reported pharmacology and acute toxicity of compounds with marihuana activity (Lowew 1946). Anticonvulsant action of marihuana-active substances was elaborated (Lowew and Goodman 1947). Antibacterial action of *C. indica* was reported in 1952 (Krejci 1952).

A case regarding development of arteritis caused by *C. indica* was described (Sterne and Ducastaing 1960). Antagonistic effect of Indian hemp on toxic-convulsant syndrome due to isonicotinic acid hydrazide was reported (Porcino 1954).

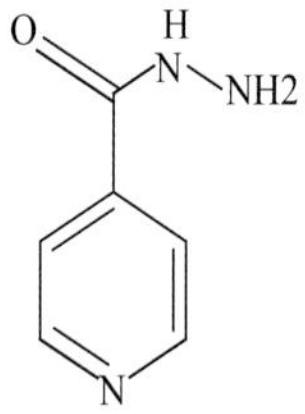

Fig. 74. Structure of isonicotinic acid hydrazide.

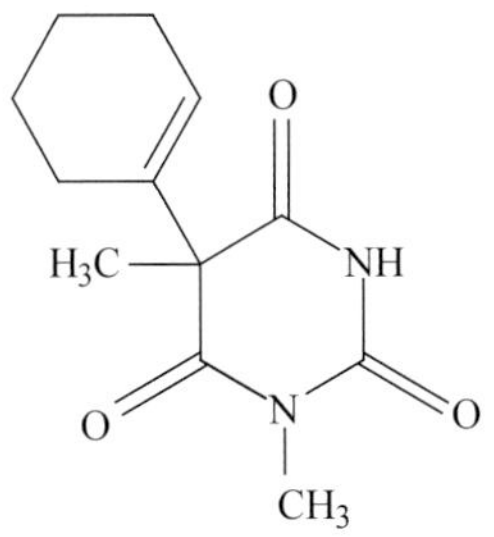

Fig. 75. Structure of hexobarbital.

A sedative and antibacterial active principle was reported from the German common hemp (Schultz and Haffner 1958; Schultz and Haffner 1959). Isolation and antibacterial properties of preparations of *C. ruderalis* growing in Ukraine was reported (Rabinovich et al. 1959). Effect of *C. indica* on hexobarbital sleeping time and tissue respiration of rat brain was reported (Bose et al. 1963).

A study reported the effects of *C. sativa* on maze performance of the rats (Carlini and Kramer 1965). Yet another study reported the effects of *C. sativa* on the fighting behaviour of mice (Santos et al. 1966). A paper dealing with the pharmacology of the hemp seed oil was published (Vieira et al. 1967). Influence of cannabis, tetrahydrocannabinol and pyrahexyl on the linguomandibular reflex of the dog was explained (Sampaio et al. 1967).

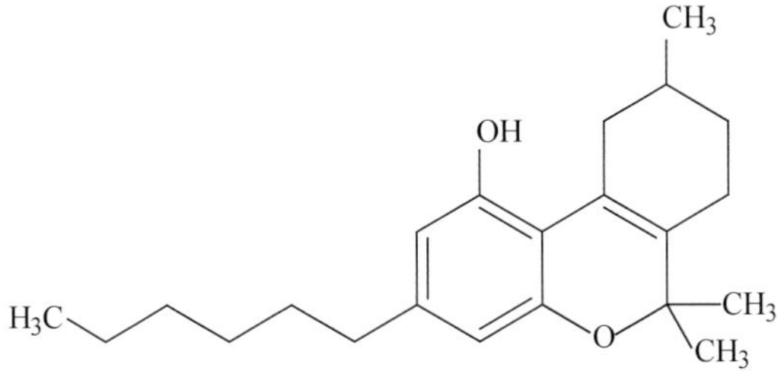

Fig. 76. Structure of pyrahexyl (a synthetic analogue of tetrahydrocannabinol).

Teratogenic activity of cannabis resin was reported (Persaud and Ellington 1968). Cannabis-induced potentiation of morphine analgesia in rats has been explained and the role of brain monoamines had been postulated (Ghosh and Bhattacharya 1979). Effects of Cannabis extract on the testicular function of the toad *Bufo andersonii* and *Presbytis entellus* have been evaluated (Dixit et al. 1977; Dixit 1980).

15.2 Detailed Pharmacological Screening

15.2.1 Cataleptogenic

Amides (feruloyltyramine and p-coumaroyltyramine) have been isolated from the ethanolic extract of cannabis seeds. The intracerebroventricular injection of both the compounds resulted in hypothermia and motor incoordination in mice. Out of the two amides, p-coumaroyltyramine exhibited cataleptogenic effect in mice (Yamamoto et al. 1991).

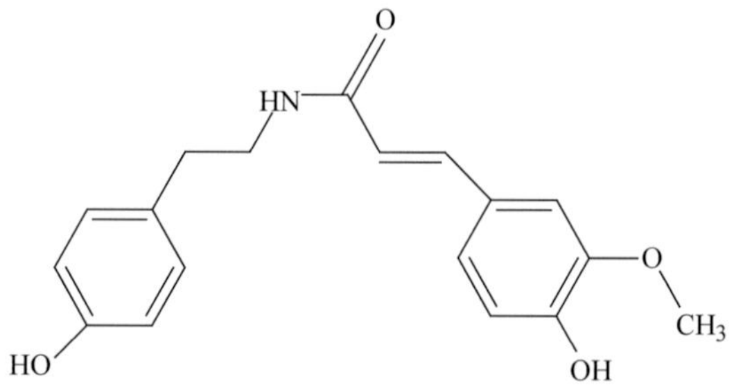

Fig. 77. Structure of feruloyltyramine.

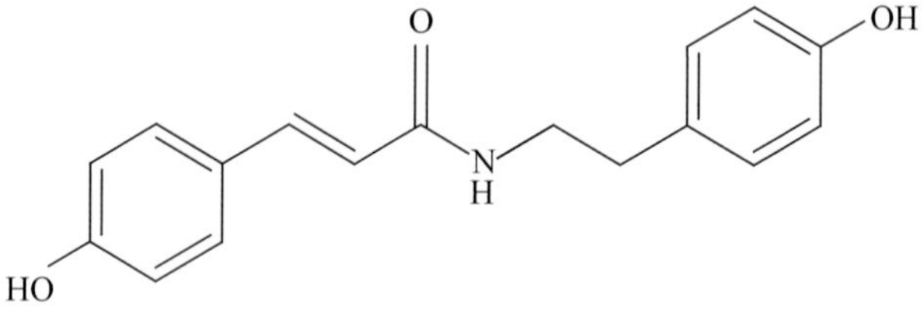

Fig. 78. Structure of p-coumaroyltyramine.

15.2.2 Trypanocidal

An aqueous extract of the seeds given at a dose of 50 mg/kg/d cured animals infected with *Trypanosoma brucei* of blood stream parasites. A couple of fractions retained the trypanocidal activity by curing mice infected with *T. brucei* (Nok et al. 1994).

15.2.3 Antimicrobial

Ethanolic extract, petroleum ether extract and the acidic fraction demonstrated significant antibacterial activity against Gram-positive and Gram-negative bacteria. The two extracts and the fraction also exhibited antifungal activity against fungi (Wasim et al. 1995).

15.2.4 Antinociceptive

An extract of cannabis in dose of 5 and 15 mg/kg (expressed as delta 9-tetrahydrocannabinol) was orally administered to rats. Naloxone (morphine antagonist), naltrexone (used in opiod dependence), and MR 1452 did not show preventive effect against the antinociceptive effect of cannabis except with high doses.

ICI 154, 129 (δ selective peptide antagonist) failed to prevent the cannabis-induced rise in nociceptive threshold except with high doses (Ferri et al. 1986).

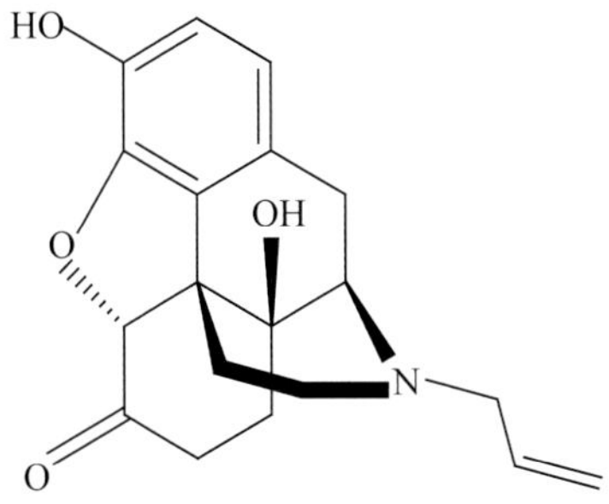

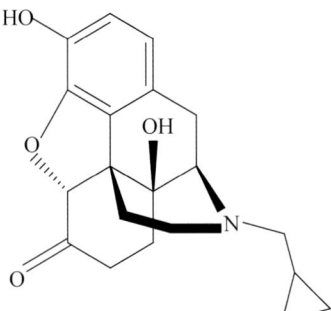

Fig. 79. Structure of naloxone.

Fig. 80. Structure of naltrexone.

15.2.5 Anticoagulant

In an *in vitro* study exploring thrombin activity, the extract based on cannabis showed high IC50 value (9.89 mg/ml) as compared to tetrahydrocannabinol (1.79 mg/ml). The results of an *in vivo* study proved in rats treated with a cannabis extract, 50% clotting times were recorded to be 1.5 and 2 fold greater than respective control groups (Coetzee et al. 2007).

15.2.6 Antihyperalgesic

In an animal model of neuropathy pain, a standardised extract of cannabis (cannabinoids, flavonoids and terpenes) demonstrated significant antinociceptive activity as compared to the single (Comelli et al. 2008).

15.2.7 Diabetic neuropathy

A series of repeated treatment in streptozotocin induced diabetic rats, with cannabis extract provided relief in allodynia. In addition, the extract caused restoration the perception of thermal pain. Above all, the cannabis extract demonstrated significant antioxidant activity in diabetic rats (Comelli et al. 2009).

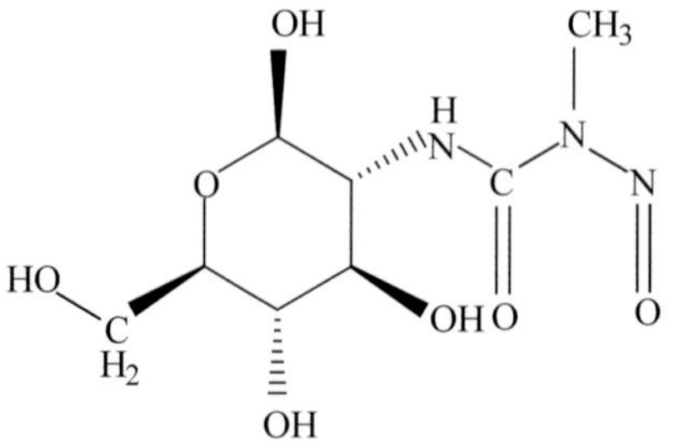

Fig. 81. Structure of streptozotocin.

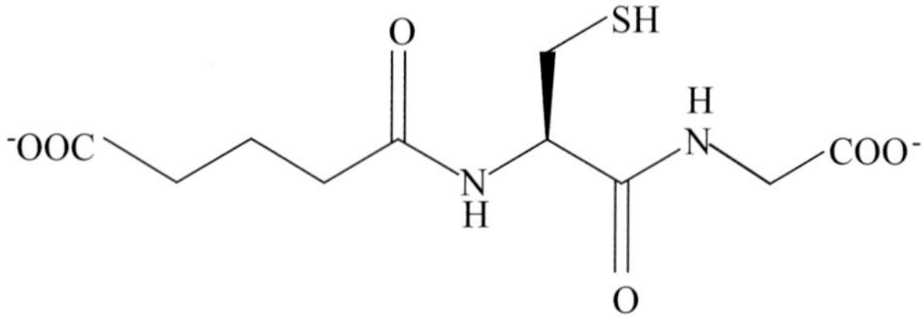

Fig. 82. Structure of glutathione.

15.2.8 Haloperidol-induced catalepsy and oxidative stress in the mice

Mice treated with cannabis extract (5, 10 or 20 mg/kg, subcutaneous route and expressed as $\Delta(9)$-THC) and haloperidol showed significant decrease in duration of catalepsy as compared to the haloperidol treatment (Abdel-Salam et al. 2012).

15.2.9 Antiobesity

Diet induced obese Wistar rats and rats were fed on standard pellets were given subcutaneous injection of cannabis extract or the vehicle for 4 weeks. In case of obese group, area under curve lowered significantly as compared to lean rats (Levendal et al. 2012).

15.2.10 Pro-resolution, protective and anti-nociceptive

Cannabis administered by intracolonic route caused reduction in the severity of hapten-induced colitis in a dose-dependent fashion. Cannabis administered by oral route cannabis reduced the severity of naproxen-induced gastric damage, and a CB1 antagonist reversed this effect (Wallace et al. 2013).

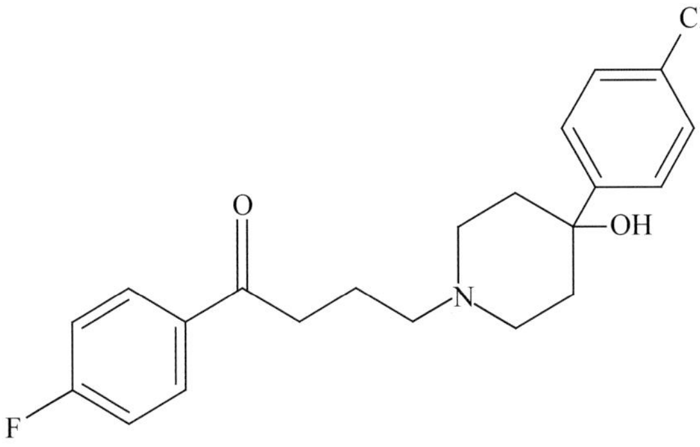

Fig. 83. Structure of haloperidol.

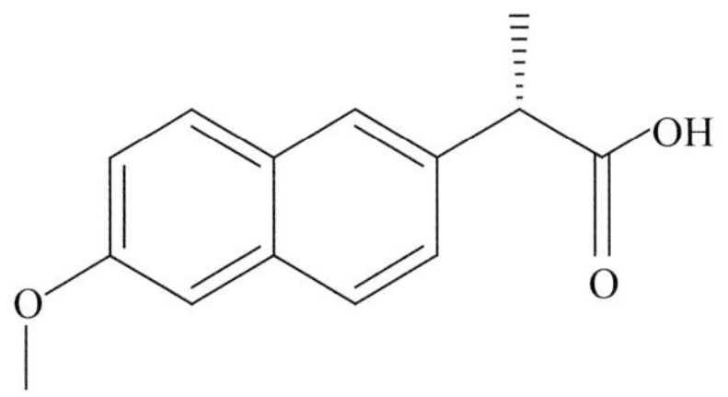

Fig. 84. Structure of naproxen.

15.2.11 Orexigenic

The first group received 0.5 mL water per day and served as vehicle group. Another group did not receive anything and served as control group. Treatment with 100 and 150 mg/kg of the cannabis extract significantly increased energy intake vs the other groups (p < 0.05). There was a significant increase in total ghrelin levels in the *C. sativa* group (Mazidi et al. 2014).

Further Reading

Abdel-Salam OM, El-Sayed El-Shamarka M, Salem NA, El-Din M Gaafar A. Effects of *Cannabis sativa* extract on haloperidol-induced catalepsy and oxidative stress in the mice. *EXCLI J.* 2012; **11:** 45–58.

Barrow B. A Case of Dysmenorrhoea in Which the Tincture of *Cannabis indica* was employed, with some observations upon that drug. *Prov Med Surg J.* 1847; **11:** 122–4.

Bose BC, Mukerji B. Observations on the physiologically active fraction of Indian hemp, *Cannabis sativa* Linn. *Indian J Med Res.* 1945; **33:** 265–70.

Bose BC, Saifi AQ, Bhagwat AW. Effect of *Cannabis indica* on hexobarbital sleeping time and tissue respiration of rat brain. *Arch Int Pharmacodyn Ther.* 1963; **141:** 520–4.

Carlini EA, Kramer C. Effects of *Cannabis sativa* (marihuana) on maze performance of the rat. *Psychopharmacologia.* 1965; **7:** 175–81.

Clendinning J. Observations on the medicinal properties of the *Cannabis sativa* of India. *Med Chir Trans.* 1843; **26:** 188–210.

Coetzee C, Levendal RA, van de Venter M, Frost CL. Anticoagulant effects of a Cannabis extract in an obese rat model. *Phytomedicine.* 2007; **14:** 333–337.

Comelli F, Bettoni I, Colleoni M, Giagnoni G, Costa B. Beneficial effects of a *Cannabis sativa* extract treatment on diabetes-induced neuropathy and oxidative stress. *Phytother Res.* 2009; **23:** 1678–1684.

Comelli F, Giagnoni G, Bettoni I, Colleoni M, Costa B. Antihyperalgesic effect of a Cannabis sativa extract in a rat model of neuropathic pain: mechanisms involved. *Phytother Res.* 2008; **22:** 1017–1024.

Dixit VP, Jain HC, Verma OP, Sharma AN. Effects of Cannabis extract on the testicular function of the toad *Bufo andersonii* Boulenger. *Indian J Exp Biol.* 1977; **15:** 555.

Dixit VP. Effects of *Cannabis sativa* on testicular function of *Presbytis entellus. Planta Med.* 1980; **41:** 288–294.

Ferri S, Cavicchini E, Romualdi P, Speroni E, Murari G. Possible mediation of catecholaminergic pathways in the antinociceptive effect of an extract of *Cannabis sativa* L. *Psychopharmacology* (Berl). 1986; **89:** 244–7.

Ghosh P, Bhattacharya SK. Cannabis induced potentiation of morphine analgesia in rate role of brain monoamines. *Indian J Med Res.* 1979; **70:** 275–280.

Inglis J. On Traumatic Tetanus and Its Treatment, with Some Remarks on the Extract of *Cannabis Indica* of Commerce. *Prov Med Surg J.* 1845; **9:** 197–200.

Krejci Z. Antibacterial action of *Cannabis indica. Lek List.* 1952; **7:** 500–3.

Levendal RA, Schumann D, Donath M, Frost CL. Cannabis exposure associated with weight reduction and β-cell protection in an obese rat model. *Phytomedicine.* 2012; **19:** 575–82.

Lowew S, Goodman LS. Anticonvulsant action of marihuana-active substances. *Fed Proc.* 1947; **6:** 352.

Lowew S. Marihuana activity of cannabinol. *Science.* 1945; **102:** 615.

Lowew S. Studies on the pharmacology and acute toxicity of compounds with marihuana activity. *J Pharmacol Exp Ther.* 1946; **88:** 154–61.

Mackenzie, S. Remarks on the value of Indian hemp in the treatment of a certain type of headache. *British Med J.* 1887; **1:** 97–98.

Mazidi M, Baghban Taraghdari S, Rezaee P, Kamgar M, Jomezadeh MR, Akbarieh Hasani O, Soukhtanloo M, Hosseini M, Gholamnezhad Z, Rakhshandeh H, Norouzy A, Esmaily H, Patterson M, Nematy M. The effect of hydroalcoholic extract of *Cannabis sativa* on appetite hormone in rat. *J Complement Integr Med.* 2014; **11:** 253–7.

Nok AJ, Ibrahim S, Arowosafe S, Longdet I, Ambrose A, Onyenekwe PC, Whong CZ. The trypanocidal effect of *Cannabis sativa* constituents in experimental animal trypanosomiasis. *Vet Hum Toxicol.* 1994; **36:** 522–4.

Persaud TV, Ellington AC. Teratogenic activity of cannabis resin. *Lancet.* 1968; **2:** 406–7.

Porcino F. Antagonistic effect of Indian hemp on toxic-convulsant syndrome due to isonicotinic acid hydrazide. *Farmaco Sci.* 1954; **9:** 278–81.

Rabinovich AS, Aizenman BIu, Zelepukha SI. Isolation and investigation of antibacterial properties of preparations from wild hemp (*Cannabis ruderalis*) growing in the Ukraine. *Mikrobiol Zh.* 1959; **21:** 40–8.

Sampaio CA, Lapa AJ, Valle JR. Influence of cannabis, tetrahydrocannabinol and pyrahexyl on the linguomandibular reflex of the dog. *J Pharm Pharmacol.* 1967; **19:** 552–4.

Santos M, Sampaio MR, Fernandes NS, Carlini EA. Effects of *Cannabis sativa* (Marihuana) on the fighting behavior of mice. *Psychopharmacologia.* 1966; **8:** 437–44.

Schultz OE, Haffner G. A sedative active principle from the German common hemp (*Cannabis sativa*). I. *Arch Pharm Ber Dtsch Pharm Ges.* 1958; **291/63:** 391–403.

Schultz OE, Haffner G. A sedative active principle from the German hemp (*Cannabis sativa*). I. *Arch Pharm Ber Dtsch Pharm Ges.* 1959; **14** b(2): 98–100.

Shaw J. On the use of the *Cannabis indica* (or Indian hemp): 1st in tetanus; 2nd in hydrophobia; 3rd in cholera—with remarks on its effects. *Madras Quart Med J.* 1843; **5:** 74–80.

Sterne J, Ducastaing C. Arteritis caused by *Cannabis indica. Arch Mal Coeur Vaiss.* 1960; **53:** 143–7.

Vieira JE, Abreu LC, Valle JR. On the pharmacology of the hemp seed oil. *Med Pharmacol Exp Int J Exp Med.* 1967; **16:** 219–24.

Wasim K, Haq I, Ashraf M. Antimicrobial studies of the leaf of *Cannabis sativa* L. *Pak J Pharm Sci.* 1995; **8:** 29–38.

Yamamoto I, Matsunaga T, Kobayashi H, Watanabe K, Yoshimura H. Analysis and pharmacotoxicity of feruloyltyramine as a new constituent and p-coumaroyltyramine in *Cannabis sativa* L. *Pharmacol Biochem Behav.* 1991; **40:** 465–9.

Chapter 16

Herbal Cannabinomimetics

16.1 *Artemisia absinthum* L. (Asteraceae)

As per data published in Nature, *Artemisia absinthum* Linn. (Absinthe) has been reported to activate the CB1 cannabinoid receptor (Nature 253: 365–356; 1975). Thujone (the active constituent in Absinthe oil) caused displacement of [3H]CP55940, a cannabinoid agonist, at concentrations above 10 microM.

The result of [35S]GTPgammaS binding assays showed that thujone was not able to stimulate G-proteins even at 0.1 mM. The compound also lacked inhibitory action on forskolin-stimulated adenylate cyclase activity in N18TG2 membranes at 1 mM. Rats administered thujone exhibited different behavioral characteristics in comparison to rats administered with levonantradol (a cannabinoid agonist) (Meschler and Howlett 1999).

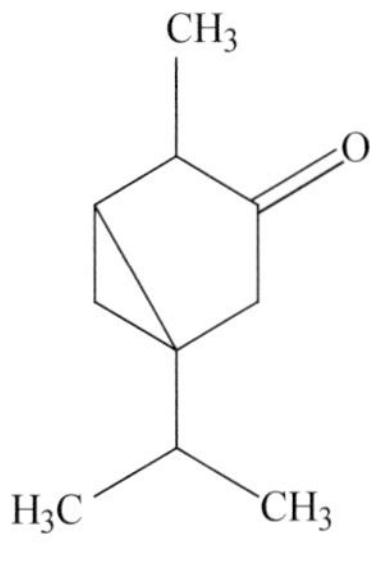

Fig. 85. Structure of thujone.

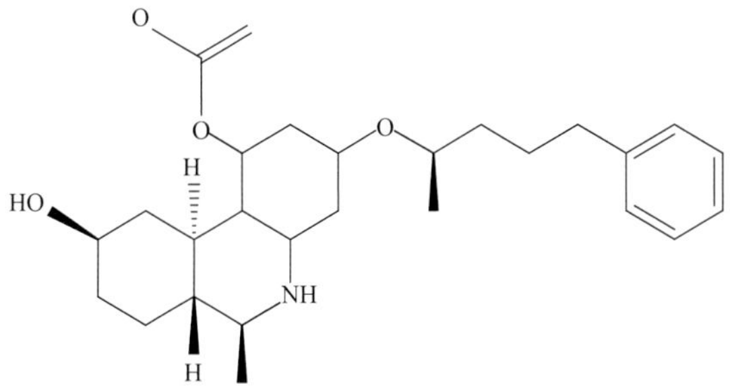

Fig. 86. Structure of levonantradol.

16.2 *Desmodium canum* Schinz and Thell. (Fabaceae)

D. incanum is commonly known as creeping beggarweed. The plant is native to Central America. Three isoflavanones isolated from the plant have with cannabinoid-like moieties (Botta et al. 2003).

16.3 *Echinacea angustifolia* L. (Asteraceae)

The plant is popularly known as the narrow-leaved purple cone flower. The alkamides found in the plant have structural similarity with anandamide. Alkamides exhibited selective affinity especially to CB2 receptors and can be considered as CB ligands (Woelkart et al. 2005).

Dodeca-2E,4E,8Z,10Z-tetraenoic acid isobutylamide and dodeca-2E,4E-dienoic acid isobutylamide bind to the CB2 receptor more strongly than the endogenous cannabinoids (Raduner et al. 2006).

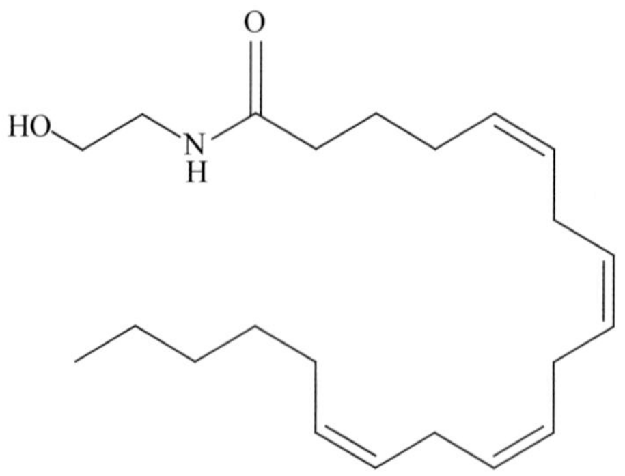

Fig. 87. Structure of anandamide.

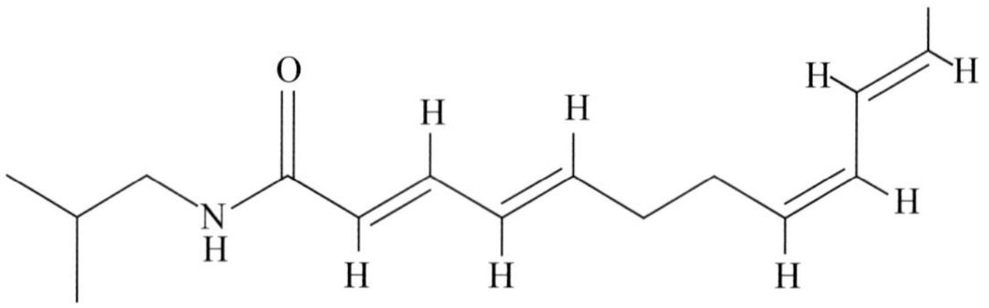

Fig. 88. Structure of Dodeca-2E,4E,8Z,10Z-tetraenoic acid isobutylamide.

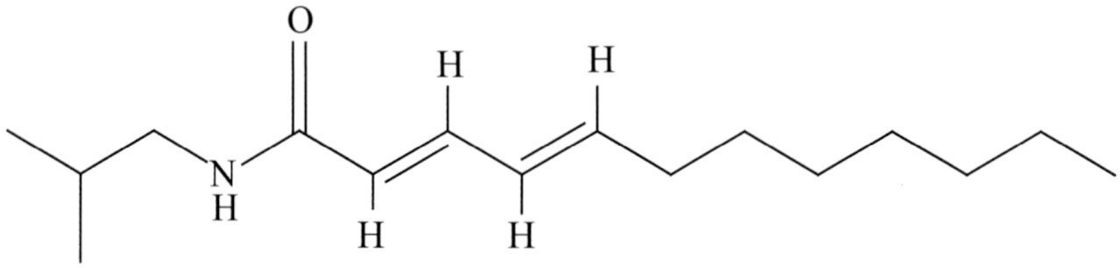

Fig. 89. Structure of Dodeca-2E,4E-dienoic acid isobutylamide.

16.4 *Echinacea purpurea* L. (Asteraceae)

The plant is popularly known as eastern purple cone flower. In rat brain membrane preparations, the bioactivity of three alkamides and nitidanin diisovalerianate from *E. purpurea* were studied in [^{35}S]GTPγS-binding experiments. The compounds showed partial as well as inverse agonist compounds for cannabinoid (CB1) receptors. This was proved by weak to moderate interactions mechanisms involved in the G-protein signaling mechanisms. However, when coadministered with arachidonyl-2'-chloroethylamide (a synthetic agonist of the cannabinoid receptor1 CB1R), the compounds inhibited the stimulation of the pure agonist. This clearly demonstrates cannabinoid receptor antagonist properties (Hohmann et al. 2011).

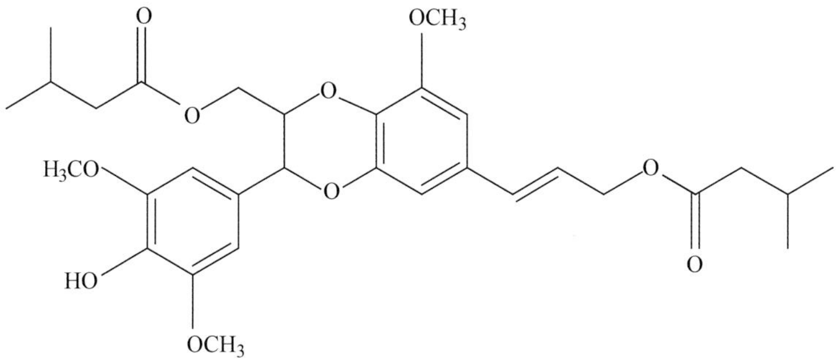

Fig. 90. Structure of nitidanin diisovalerianate.

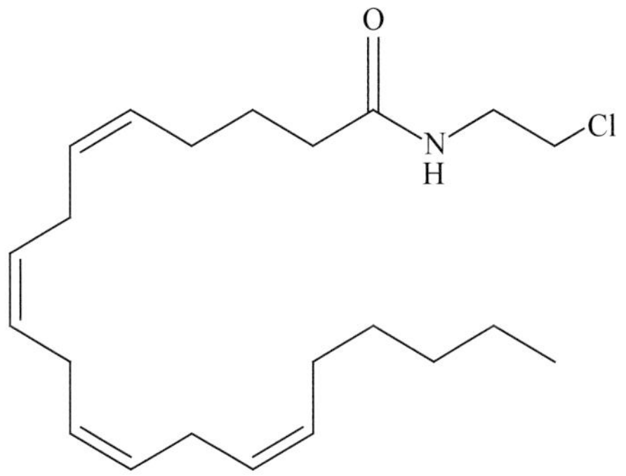

Fig. 91. Structure of arachidonyl-2'-chloroethylamide.

16.5 *Helichrysum umbraculigerum* Less. (Asteraceae)

H. umbraculigerum is commonly known as the woolly umbrella *Helichrysum* and widely distributed in South Africa. In traditional medicine the plant is used in the treatment of depression and mood disorders. It is reported to accumulate large amount of cannabigerol (Bohlman and Hoffmann 1979). Recently, amorfrutin-type phytocannabinoids have been reported from *H. umbraculigerum* (Pollastro et al. 2017).

16.6 *Radula marginata* Taylor

R. marginata is a liverwort endemic to New Zealand. Cannabinoid type bibenzyl compounds including perrottetinenic acid, perrottetinene and isoperrottetin A have been reported (Toyota et al. 2002). The chemical structure of perrottetinene resembles with THC. Perrottetinene is postulated to be an active cannabinoid agonist, however detailed pharmacological investigations are missing.

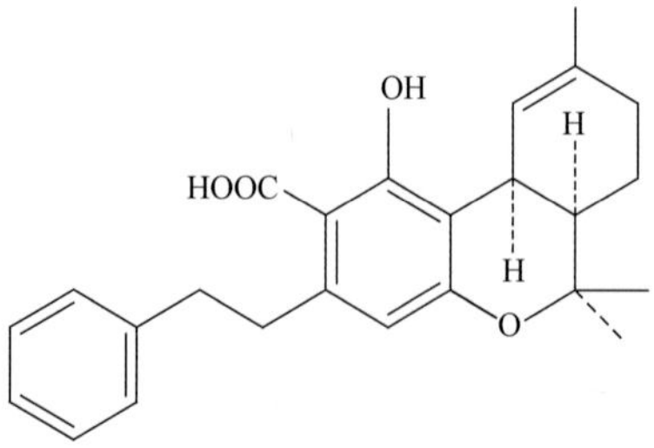

Fig. 92. Structure of perrottetinenic acid.

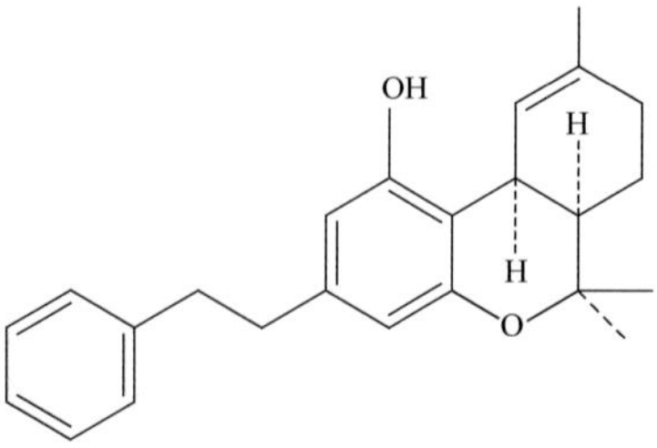

Fig. 93. Structure of perrottetinene.

16.7 Rutamarin from *Ruta graveolens*

Rutamarin in *Ruta graveolens* has micromolar affinity for CB_2 (Rollinger et al. 2009).

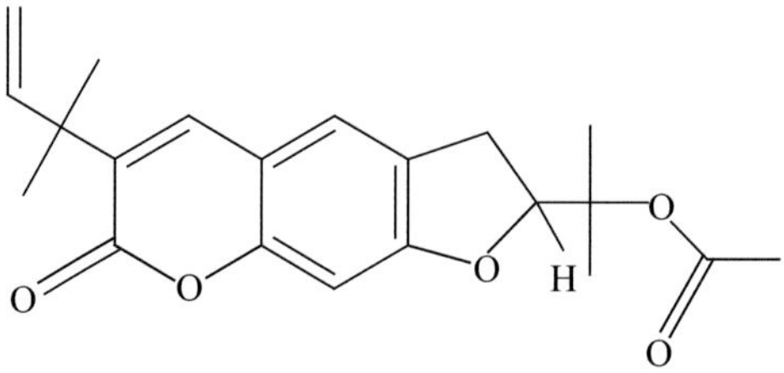

Fig. 94. Structure of rutamarin.

16.8 (*E*)-β-caryophyllene

(*E*)-β-caryophyllene is a major constituent in *Cannabis*. The constituent has selective binding to cannabinoid receptor type 2. Upon binding, the constituent has inhibitory action on adenylate cyclase. This results in transition on calcium inside the cell.

(*E*)-β-caryophyllene is reported to inhibit lipopolysaccharide-induced proinflammatory cytokine expression in the peripheral blood stream. The terpene by oral route (5 mg/kg dose) causes significant reduction of carrageenan-induced inflammation in wild-type mice. The finding clearly prove that (*E*)-β-caryophyllene has cannabimimetic effect *in vivo* (Gertsch et al. 2008).

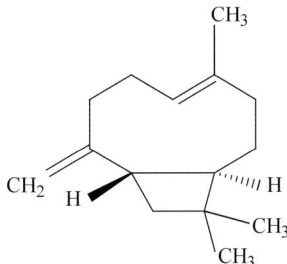

Fig. 95. Structure of β-caryophyllene.

16.9 Falcarinol (panaxynol, carotatoxin) from *Seseli praecox*

Falcarinol is a fatty alcohol found in members of Apiaceae including carrots, parsley, and celery, and in *Panax ginseng*. Falcarinol has been isolated from *Seseli praecox* (Gamisans) Gamisans (Apiaceae), endemic to Sardinia.

Falcarinol exhibited binding affinity to both human cannabinoid receptors. However, falcarinol selectively alkylates the anandamide binding site in the CB(1) receptor (K(i) = 594 nM). Falcarinol inhibits the effects of anandamide on tissue necrosis factor-alpha stimulated keratinocytes (Leonti et al. 2010).

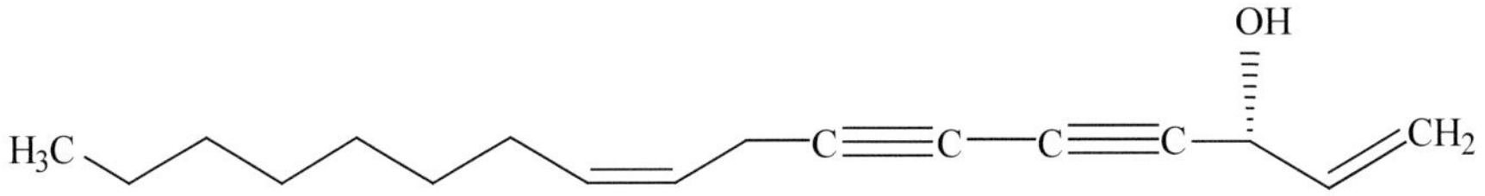

Fig. 96. Structure of (R)-(–)-Falcarinol.

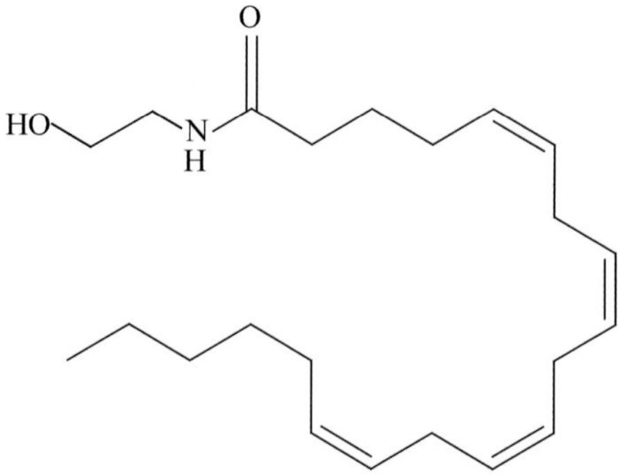

Fig. 97. Structure of anandamide (*N*-arachidonoylethanolamine).

16.10 *Lyngbya majuscula* Harvey ex Gomont (Oscillatoriaceae)

Cyclopropyl-containing metabolites (grenadadiene, debromogrenadiene and grenadamide) have been isolated from *L. majuscula* collected from Grenada. Grenadamide demonstrated modest brine shrimp toxicity (LD50 = 5 microg/mL). The compound also showed cannabinoid receptor binding activity (Ki = 4.7 microM) (Sitachitta and Gerwick 1998).

Nuclear magnetic resonance-guided fractionation of independent collections of *Lyngbya majuscula* (collected from Papua New Guinea) and *Oscillatoria* sp. (collected from Panama) resulted in isolation of the new lipids (serinolamide A and propenediester). Serinolamide A showed a moderate agonist activity and selectivity for the CB1 cannabinoid receptor (Ki = 1.3 µM, > 5-fold) (Gutierrez et al. 2011).

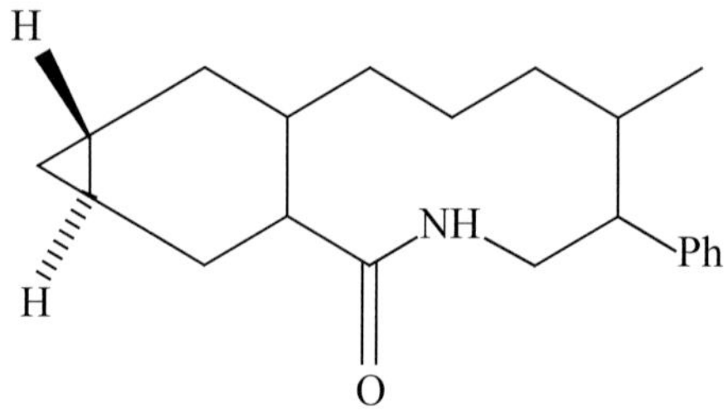

Fig. 98. Structure of grenadamide.

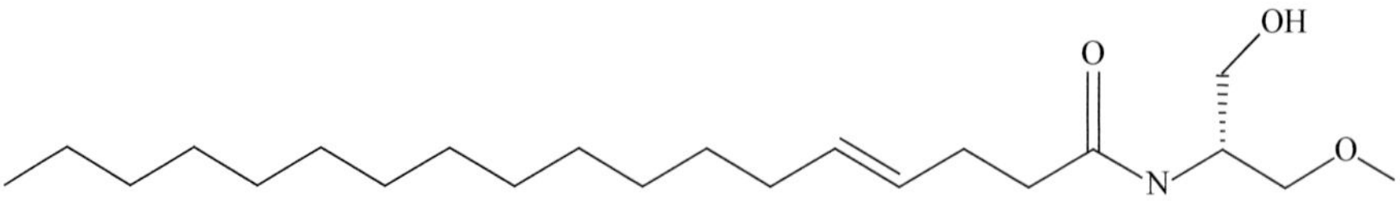

Fig. 99. Structure of serinolamide A.

16.11 Pristimerin and Euphol

Two naturally occurring terpenoids, pristimerin and euphol have been reported to inhibit monoacylglycerol lipase (MAG lipase) activity with high potency (King et al. 2009).

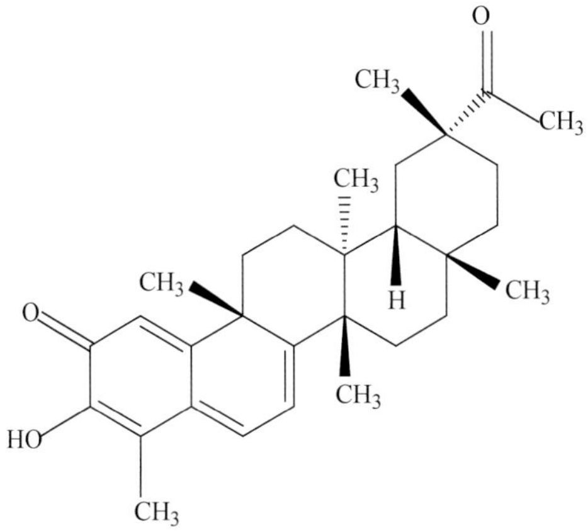

Fig. 100. Structure of pristimerin.

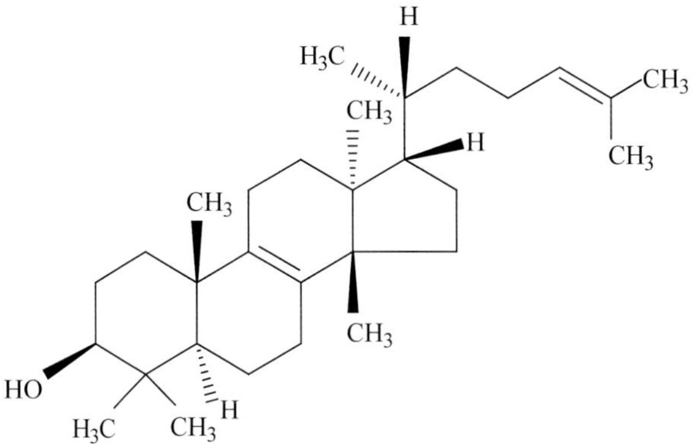

Fig. 101. Structure of euphol.

16.12 Mooreamide A

Mooreamide A isolated from *Moorea bouillonii* Hoffmann & Demoulin (Oscillatoriophycideae) has been reported to have potent and selective ligand binding activity to CB(1) (K(1) = 0.47 μM) versus CB(2) (K(1) > 25 μM) (Mevers et al. 2014).

Further Reading

Botta B, Gacs-Baitz E, Vinciguerra V, Delle Monache G. Three isoflavanones with cannabinoid-like moieties from *Desmodium canum*. *Phytochemistry*. 2003; **64:** 599–602.

Gertsch J, Leonti M, Raduner S, Racz I, Chen JZ, Xie XQ, Altmann KH, Karsak M, Zimmer A. Beta-caryophyllene is a dietary cannabinoid. *Proc Natl Acad Sci USA*. 2008; **105:** 9099–104.

Gutierrez M, Pereira AR, Debonsi HM, Ligresti A, Di Marzo V et al. Cannabinomimetic lipid from a marine cyanobacterium. *J Nat Prod*. 2011; **74:** 2313–2317.

Hohmann J, Rédei D, Forgo P, Szabó P, Freund TF, Haller J, Bojnik E, Benyhe S. Alkamides and a neolignan from Echinacea purpurea roots and the interaction of alkamides with G-protein-coupled cannabinoid receptors. *Phytochemistry*. 2011; **72:** 1848–53.

http://cannabisculture.hanf.ws/tag/wormwood/.

Leonti M, Casu L, Raduner S, Cottiglia F, Floris C, Altmann KH, Gertsch J. Falcarinol is a covalent cannabinoid CB1 receptor antagonist and induces pro-allergic effects in skin. *Biochem Pharmacol*. 2010; **79:** 1815–26.

Meschler JP, Howlett AC. Thujone exhibits low affinity for cannabinoid receptors but fails to evoke cannabimimetic responses. *Pharmacol Biochem Behav*. 1999; **62:** 473–480.

Mevers E, Matainaho T, Allara' M, Di Marzo V, Gerwick WH. Mooreamide A: a cannabinomimetic lipid from the marine cyanobacterium *Moorea bouillonii*. *Lipids*. 2014; **49:** 1127–32.

Raduner S, Majewska A, Chen J-Z, Xie X-Q, Hamon J, Faller B, Altmann K-H, Gertsc J. Alkylamides from Echinacea are a new class of cannabinomimetics. *Biol Chem*. 2006; **281:** 14192–14206.

Rollinger JM, Schuster D, Danzl B, Schwaiger S, Markt P et al. *In silico* target fishing for rationalized ligand discovery exemplified on constituents of *Ruta graveolens*. *Planta Med*. 2009; **75:** 195–204.

Sitachitta N, Gerwick WH. Grenadadiene and grenadamide, cyclopropyl-containing fatty acid metabolites from the marine cyanobacterium *Lyngbya majuscula*. *J Nat Prod*. 1998; **61:** 681–684.

Toyota M, Shimamura T, Ishii H, Renner M, Braggins J, Asakawa Y. New Bibenzyl Cannabinoid from the New Zealand Liverwort *Radula marginata*. *Chem Pharm Bull*. 2002; **50:** 1390–1392.

Toyota M, Tomohide K, Asakawa Y. Bibenzyl cannabinoid and bisbibenzyl derivative from the liverwort *Radula perrottetii. Phytochemistry.* 1994; **37:** 859.

Woelkart K, Xu W, Pei Y, Makriyannis A, Picone RP, Bauer R. The endocannabinoid system as a target for alkamides from *Echinacea angustifolia* roots. *Planta Med*. 2005; **71:** 701–5.

Chapter 17

Cannabis Oil

17.1 Chemistry

17.1.1 Essential fatty acids

The edible oil contains approximately 80% essential fatty acids (Oomaha et al. 2002; Leizer et al. 2002; Borhade 2013). Hempseed oil is a rich and balanced source of omega-6 and omega-3 polyunsaturated fatty acids. The constituents of the oil are as follows (Deferne and Pate 1996):

Linolenic acid: Hemp oil contains 60% linolenic acid (Fig. 102).
α-Linolenic acid: Hemp oil contains 1–4% α-linolenic acid (Fig. 103).
γ-Linolenic acid: Hemp oil contains 60% linolenic acid (Fig. 104).
Stearidonic acid: Hemp oil contains 0–2% stearidonic acid (Fig. 105).
Oleic acid: Hemp oil contains 11% oleic acid (Fig. 106).

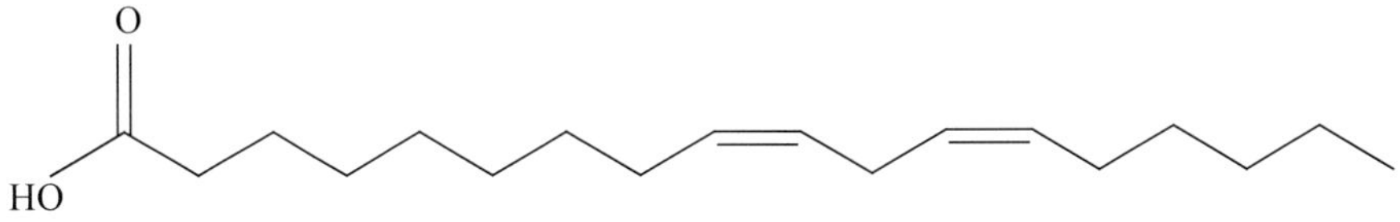

Fig. 102. Structure of linolenic acid.

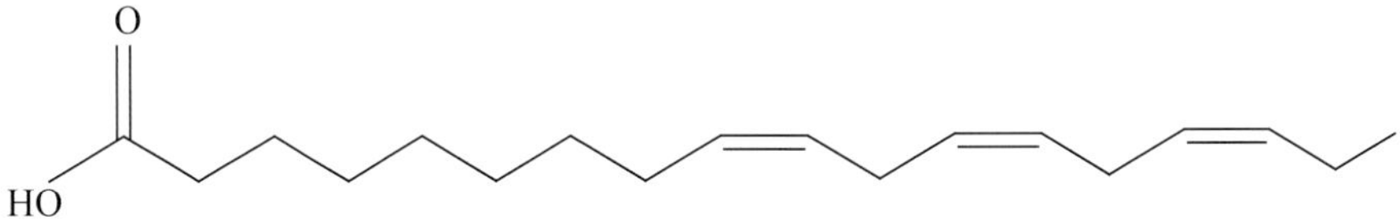

Fig. 103. Structure of α-Linolenic acid.

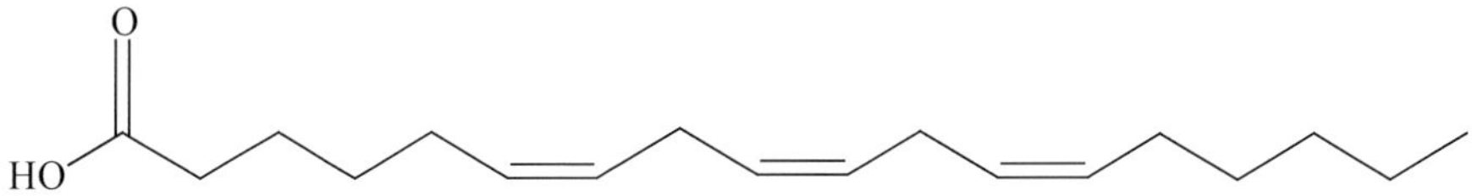

Fig. 104. Structure of γ-Linolenic acid.

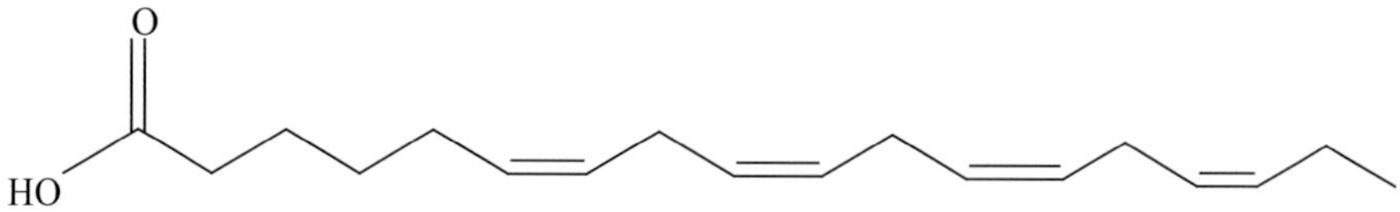

Fig. 105. Structure of stearidonic acid.

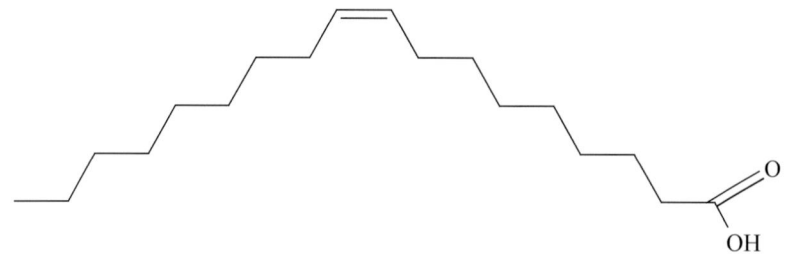

Fig. 106. Structure of oleic acid.

17.1.2 Tocopherols

Polyphenolic Compounds and Antioxidant Activity of Cold-Pressed Seed Oil from Finola Cultivar of Cannabis sativa has been reported (Smeriglio et al. 2016).

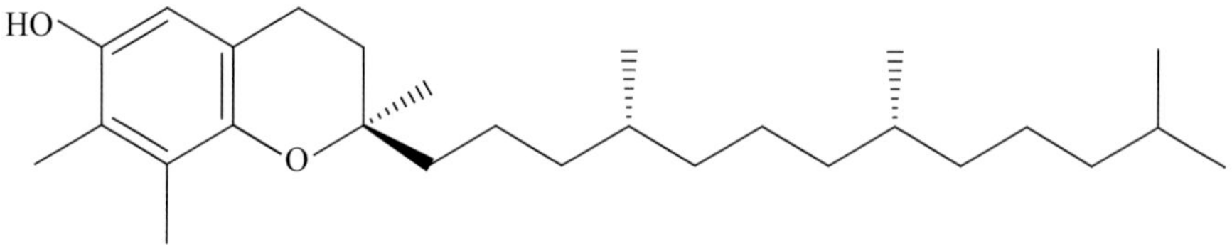

Fig. 107. Structure of γ-tocopherol.

Microwave treatment resulted in improvement of the yield of oil. It increased the amount of carotenoid and other pigments. The treatment, however, decreased *p*-anisidine value. As far as tocopherols are concerned, β-tocopherol concentrations was found to be increased. γ-tocopherol, and fatty acid composition of the oil was found to be unaffected by microwave treatment of hempseed.

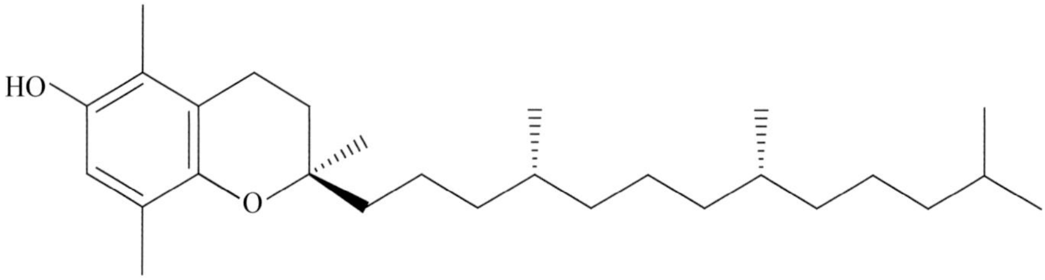

Fig. 108. Structure of β-tocopherol.

In China, eight cultivars of hempseed were collected from different locations. The concentration of unsaturated fatty acids was reported to exceed 90%. The range of polyunsaturated fatty acids was from 76.26%–82.75%. Linoleic acid and α-linolenic acid (3:1 ratio) were the main components of the polyunsaturated fatty acids. The concentration of γ-tocopherol was 28.23 mg/100 g of hempseed oil (Chen et al. 2010).

17.1.3 Phytosterols

Phytosterols has been reported in red oil extract of cannabis (Fenselau and Hermann 1972).

The unsaponifiable fraction contains β-sitosterol (1905.00 ± 59.27 mg/kg of oil), campesterol (505.69 ± 32.04 mg/kg of oil), phytol (167.59 ± 1.81 mg/kg of oil), cycloartenol (90.55 ± 3.44 mg/kg of oil), and γ-tocopherol (73.38 ± 2.86 mg/100 g of oil) (Montserrat-de la Paz et al. 2014).

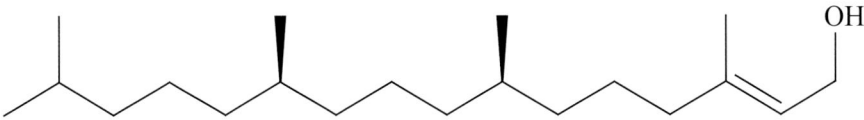

Fig. 109. Structure of phytol.

17.1.4 Lignanamides

Cannabisin M, cannabisin N, cannabisin O and 3,3'-demethyl-heliotropamide have been isolated from the seeds (Yan et al. 2015).

17.1.5 Essential oil

Essential oil from fresh *C. sativa* has been used in identification of the plant (Martin et al. 1961). Pharmacological evaluation of volatile oil have been investigated (Segelman et al. 1974). The bacteriostatic activity of hemp essential oil has been reported (Fournier et al. 1978).

Drying of the plant material had no effect on the qualitative composition of the oil and did not affect the ability of individuals familiar with marijuana smell to recognize the odor (Ross and ElSohly 1996). The essential oils of Carmagnola, Fibranova and Futura strain significantly inhibits the microbial growth, to an extent depending on variety and sowing time (Nissen et al. 2010).

17.2 Preclinical Pharmacology

Antioxidant: The cold-pressed oil from Finola cultivar of *C. sativa* has high antioxidative activity (146.76 mmol of TE/100 g oil). The oil had inhibitory activity on β-carotene bleaching. Further, it resulted in quenching of peroxyl radical *in vitro*. Reactivity towards 2,2'-azino-bis (3-ethylbenzothiazoline-6-sulfonic acid) radical cation and ferric-reducing antioxidant power values were 695.2 μmol of TE/100 g oil and 3690.6 μmol of TE/100 g oil respectively (Shelef et al. 2016).

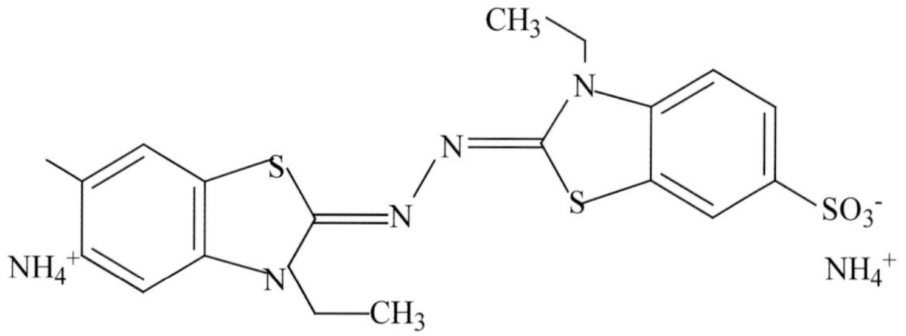

Fig. 110. Structure of 2,2'-azino-bis (3-ethylbenzothiazoline-6-sulfonic acid).

17.3 Clinical Studies

17.3.1 Dementia

An open label study lasting for 28 days, recruited eleven patients diagnosed with Alzheimer's Dementia. A significant reduction in Clinical Global Impression severity score was recorded (6.5 to 5.7; $p < 0.01$). Similarly a significant reduction in the neuropsychiatric inventory was noticed (44.4 to 12.8; $p < 0.01$) (Shelef et al. 2016).

17.3.2 Atopic dermatitis

A 20-week randomized, single blind study compared efficacy of olive oil and dietary hempseed oil in the treatment of atopic dermatitis. Dietary hempseed oil was reported to cause significant reduction in the clinical symptoms of the disease (Callaway et al. 2005).

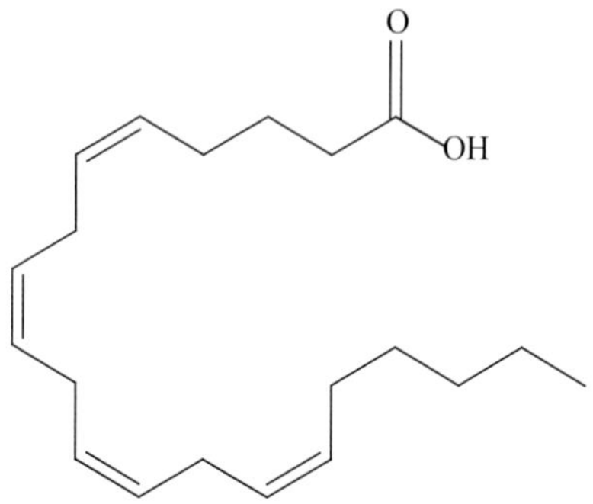

Fig. 111. Structure of arachidonic acid.

17.3.3 A comparative study on cardiovascular health in healthy volunteers

In a double blinded, placebo controlled clinical trial lasting for 3 months, 86 healthy participants were divided into four groups.

   1st group received 1 gram capsule of placebo, twice-a-day,
   2nd group received 1 gram capsule of fish oil, twice-a-day,

3rd group received 1 gram capsule of flaxseed oil, twice-a-day,
4th group received 1 gram capsule of hempseed oil, twice-a-day.

The concentration of the fatty acids of eicosapentanoic acid and docosahexaenoic acid increased significantly in the plasma with the intake of fish oil. A gradual increase in the levels of α-linolenic acid was noticed after administration of flaxseed oil. Above all, hempseed oil was not able to cause significant increase in concentration of the fatty acids in the plasma (Kaul et al. 2008).

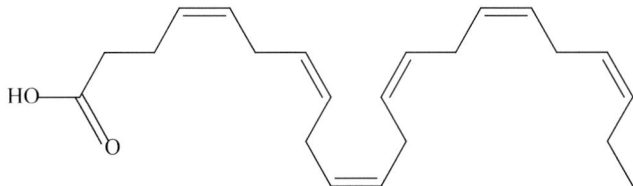

Fig. 112. Structure of docosahexaenoic acid.

Fig. 113. Structure of eicosapentanoic acid.

Further Reading

Borhade SS. Chemical composition and characterization of hemp (*Cannabis sativa*) seed oil and essential fatty acids by HPLC Method. *Arch Appl Sci Res*. 2013; **5:** 5–8.

Callaway J, Schwab U, Harvima I, Halonen P, Mykkänen O, Hyvönen P, Järvinen T. Efficacy of dietary hempseed oil in patients with atopic dermatitis. *J Dermatolog Treat*. 2005; **16:** 87–94.

Chen T, He J, Zhang J, Zhang H, Qian P, Hao J, Li L. Analytical characterization of Hempseed (seed of *Cannabis sativa* L.) oil from eight regions in China. *J Diet Suppl*. 2010; **7:** 117–29.

Deferne JL, Pate DW. Hemp seed oil: A source of valuable essential fatty acids. *J Int Hemp Assoc*. 1996; **1:** 4–7.

Fenselau C, Hermann G. Identification of phytosterols in red oil extract of cannabis. *J Forensic Sci*. 1972; **17:** 309–12.

Fournier G, Paris MR, Fourniat MC, Quero AM. Bacteriostatic activity of *Cannabis sativa* L. essential oil. *Annales Pharmaceutiques Françaises*. 1978; **36:** 603–6.

Kaul N, Kreml R, Austria JA, Richard MN, Edel AL, Dibrov E, Hirono S, Zettler ME, Pierce GN. A comparison of fish oil, flaxseed oil and hempseed oil supplementation on selected parameters of cardiovascular health in healthy volunteers. *J Am Coll Nutr*. 2008; **27:** 51–8.

Leizer C, Ribnicky D, Poulev A, Dushenkov S, Raskin I. The composition of hemp seed oil and its potential as an important source of nutrition. *J Nutr Funct Med Foods*. 2000; **2:** 35–52.

Malingré T, Herndriks H, Battermann S, Bos R, Visser J. The essential oil of *Cannabis sativa*. *Planta Med*. 1975; **28:** 56–61.

Martin L, Smith DM, Farmilo CG. Essential oil from fresh *Cannabis sativa* and its use in identification. *Nature*. 1961; 19; **191:** 774–6.

Montserrat-de la Paz S, Marín-Aguilar F, García-Giménez MD, Fernández-Arche MA. Hemp (*Cannabis sativa* L.) seed oil: analytical and phytochemical characterization of the unsaponifiable fraction. *J Agric Food Chem*. 2014; **62**: 1105–10.

Nigam MC, Handa KL, Nigam IC, Levi L. Essential oils and their constituents XXIX: the essential oil of marihuana: composition of genuine Indian *Cannabis sativa* L. *Can J Chem*. 1965; **43**: 3372–3376.

Nissen L, Zatta A, Stefanini I, Grandi S, Sgorbati B, Biavati B, Monti A. Characterization and antimicrobial activity of essential oils of industrial hemp varieties (*Cannabis sativa* L.). *Fitoterapia*. 2010; **81**: 413–9.

Oomaha BD, Bussonb M, Godfreya DV, Drovera CG. Characteristics of hemp (*Cannabis sativa* L.) seed oil. *Food Chem*. 2002; **76**: 33–43.

Ross SA, ElSohly M. The volatile oil composition of fresh and air-dried buds of *Cannabis sativa*. *J Natl Prod*. 1996; **59**: 49–51.

Segelman AB, Sofia RD, Segelman FP, Harakal JJ, Knobloch LC. *Cannabis sativa* L. (marijuana). V. Pharmacological evaluation of marijuana aqueous extract and volatile oil. *J Pharm Sci*. 1974; **63**: 962–4.

Shelef A, Barak Y, Berger U, Paleacu D, Tadger S, Plopsky I, Baruch Y. Safety and Efficacy of medical cannabis oil for behavioral and psychological symptoms of dementia: An-open label, Add-On, pilot study. *J Alzheimers Dis*. 2016; **51**: 15–9.

Smeriglio A, Galati EM, Monforte MT, Lanuzza F, D'Angelo V, Circosta C. Polyphenolic Compounds and Antioxidant Activity of Cold-Pressed Seed Oil from Finola Cultivar of *Cannabis sativa* L. *Phytother Res*. 2016; **30**: 1298–307.

Yan X, Tang J, dos Santos Passos C, Nurisso A, Simões-Pires CA, Ji M, Lou H, Fan P. Characterization of lignanamides from hemp (*Cannabis sativa* L.) seed and their antioxidant and acetylcholinesterase inhibitory activities. *J Agric Food Chem*. 2015; **63**: 10611–9.

Chapter 18

Wild Cannabis—*Leonotis leonurus* (L.) R.Br.

18.1 Common Name

Lion's Ear, Lion's Tail, Wild Dagga, Dacha, Daggha (Africa), Wild Hemp, Minaret Flower, Flor de Mundo, Mota (Mexico) and wild dagga (this name links the plant with cannabis).

18.2 Habitat

Native to South Africa and South America.

18.3 Botany

L. leonurus is a robust shrub which grows upto 2–3 m tall and 1.5 m wide. Stems are velvety and woody at the base. The leaves are long, narrow, rough above, velvety below, with serrate edges. The wild dagga flowers profusely in autumn with its characteristic bright orange flowers carried in compact clusters in whorls along the flower stalk. Apricot and creamy white flowered forms are also found.

18.4 Phytochemistry

Alkaloid: leonurine (Hayashi 1962) and diterpenoids: leonurenones A–C, 14α-hydroxy-9α,13α-epoxylabd-5(6)-en-7-on-16,15-olide and 13ξ-hydroxylabd-5(6),8(9)-dien-7-on-16,15-olide (He et al. 2012; Narukawa et al. 2015) and luteolin 7-*O*-β-glucoside and luteolin (Agnihotri et al. 2009).

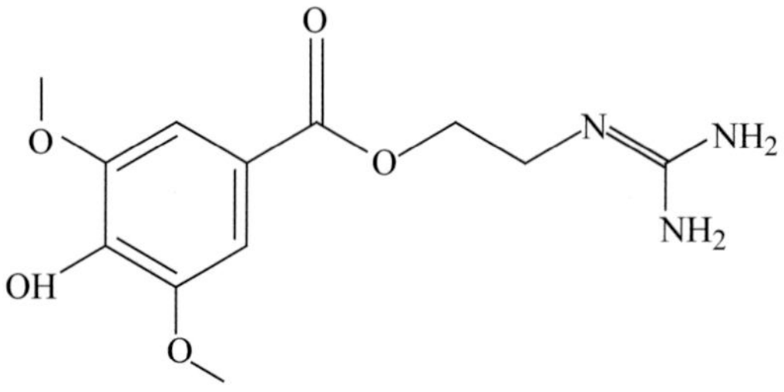

Fig. 114. Structure of leonurine.

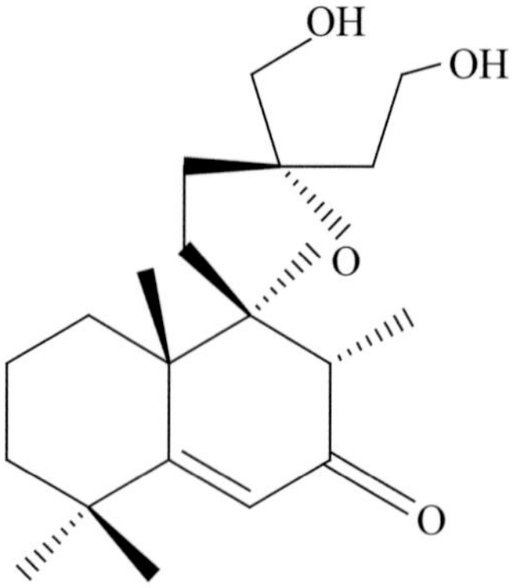

Fig. 115. Structure of leonurenone A.

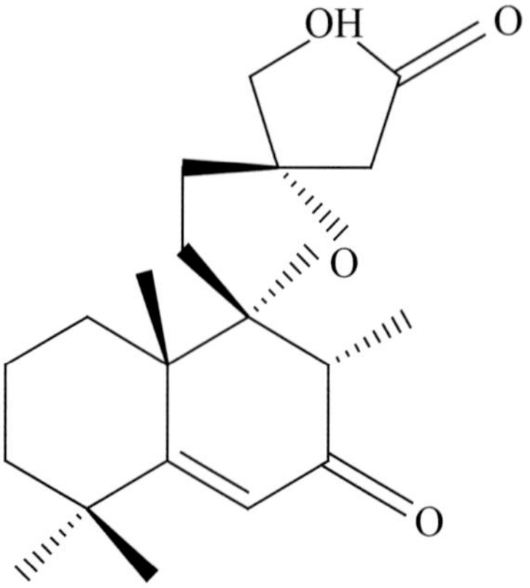

Fig. 116. Structure of leonurenone B.

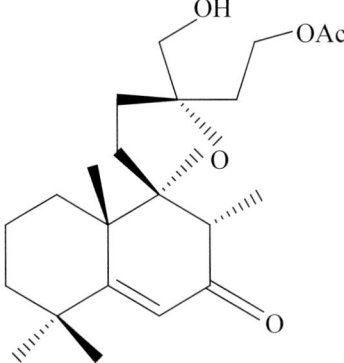

Fig. 117. Structure of leonurenone C.

18.5 Therapeutics

L. leonurus is used in the treatment of fever, piles (haemorrhoids), eczema (dermatitis), skin rashes, carbuncle, prurits, muscular cramps, cephalgia, epilepsy, chest infections, constipation, and snake bite. *L. leonurus* is reported to have mild hallucinogenic effect. The hallucinogenic effect is seen in dried buds or leaves that are used for the purpose of smoking. The dried leaves and flowers are smoked to relieve epilepsy (Nsuala et al. 2015).

In some cultures, people use *L. leonurus* with cannabis for smoking. *L. leonurus* has been used as a substitute for marijuana. In Mexico, *L. leonurus* is known as *flor de mundo* and *mota* and is used as a substitute for Cannabis. In South Africa, some cultures use *L. leonurus* in the treatment of arthritis, inflammation and type-2 diabetes mellitus.

18.6 Pharmacology of *L. leonurus*

Antinociceptive, anti-inflammatory and antidiabetic: Aqueous extract of the leaves of *L. leonurus* in a dose of 50–800 mg/kg (intra-peritoneally) resulted in antinociceptive effect. This effect was exerted against chemically and thermally induced nociceptive pain stimuli in mice. The effect was dose-dependent. With the same dose, the extract significantly showed anti-inflammatory effect against fresh egg albumin-induced paw edema. In rats, aqueous extract showed significant hypoglycemic effects (Ojewole 2005).

Anti-epileptic: Aqueous extract of *L. leonurus* (doses of 200 and 400 mg/kg) resulted in delayed pentylenetetrazole (90 mg/kg)-induced tonic seizures. The water extract in same doses resulted in delay of onset of tonic seizures induced by picrotoxin (8 mg/kg) and N-methyl-DL-aspartic acid (400 mg/kg) (Bienvenu et al. 2002).

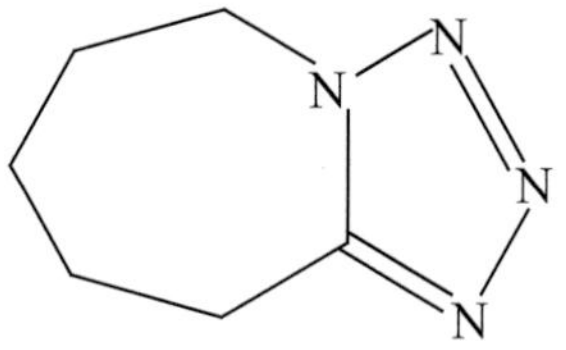

Fig. 118. Structure of pentylenetetrazole.

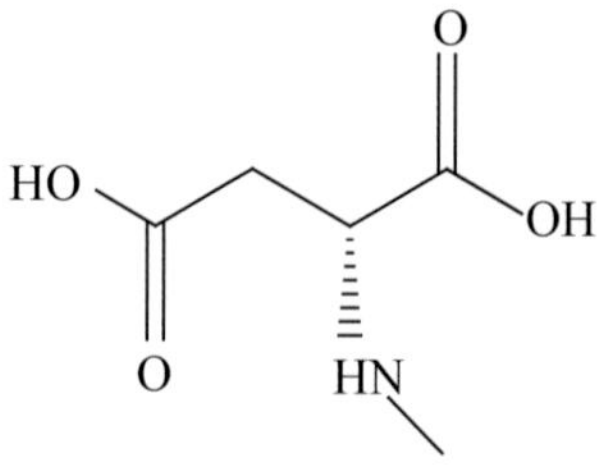

Fig. 119. Structure of N-methyl-DL-aspartic acid.

18.7 Pharmacology of Leonurine

Neuroprotective: An investigatory study was undertaken for evaluating therapeutic effect of leonurine on ischemic stroke using middle cerebral artery occlusion method. The animals were pretreated with leonurine orally for one week followed by surgery. Pretreatment with leonurine reduced the volume of the infarct. The alkaloid resulted in improvement in neurological deficit in stroke groups. The alkaloid also showed antioxidant activity and decreased levels of malondialdehyde (Fig. 120) (Loh et al. 2010).

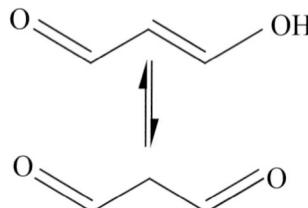

Fig. 120. Structure of malondialdehyde.

Cardioprotective: Pretreatment with leonurine attenuated lipopolysaccharide-induced mRNA expression of intercellular adhesion molecule-1, vascular cell adhesion molecule-1, E-selectin, and monocyte chemoattractant protein-1 in a dose dependent fashion. Inhibition of free radicals and NF-κB signalling pathways seems to be the underlying mechanism of anti-inflammatory activity of the alkaloid (Liua et al. 2012).

Anti-tumour: A study investigated the anti-tumour activity of leonurine on human non-small cell lung cancer H292 cells. The alkaloid significantly inhibited the proliferation of H292 cells in a time and dose-dependent manner. Leonurine also induced G0/G1 cell-cycle arrest (Mao et al. 2015).

Nephroprotective: In a study based on acute kidney failure evoked by lipopolysaccharide, leonurine suppressed NF-κB activation and inhibited pro-inflammatory cytokine production via decreasing cellular reactive oxygen species production (Xu et al. 2014).

Further Reading

Agnihotri VK, ElSohly HN, Smillie TJ, Khan IA, Walker LA. Constituents of *Leonotis leonurus* flowering tops. *Phytochem Lett*. 2009; **2:** 103–105.

Bienvenu E, Amabeoku GJ, Eagles PK, Scott G, Springfield EP. Anticonvulsant activity of aqueous extract of *Leonotis leonurus*. *Phytomedicine*. 2002; **9:** 217–223.

Hayashi Y. Studies on the ingredients of *Leonurus sibiricus* L. II. Structure of leonurine (2) *Yakugaku Zasshi*. 1962; **82:** 1025–1027.

He F, Lindqvist C, Hardinga WW. Leonurenones A–C: Labdane diterpenes from *Leonotis leonurus*. *Phytochemistry*. 2012; **83:** 168–172.

Liua XH, Pana LL, Yanga HB, Gonga QH, Zhua YZ. Leonurine attenuates lipopolysaccharide-induced inflammatory responses in human endothelial cells: Involvement of reactive oxygen species and NF-κB pathways. *European J Pharmacol*. 2012; **680:** 108–114.

Loh KP, Qi J, Tan BK, Liu XH, Wei BG, Zhu YZ. Leonurine protects middle cerebral artery occluded rats through antioxidant effect and regulation of mitochondrial function. *Stroke*. 2010; **41:** 2661–8.

Mao F, Zhang L, Cai MH, Guo H, Yuan HH. Leonurine hydrochloride induces apoptosis of H292 lung cancer cell by a mitochondria-dependent pathway. *Pharm Biol*. 2015; **53:** 1684–90.

Narukawa Y, Komori M, Niimura A, Noguchi H, Kiuchi F. Two new diterpenoids from *Leonotis leonurus* R. Br. *J Nat Med*. 2015; **69:** 130–4.

Nsuala BN, Enslin G, Viljoen A. "Wild cannabis": A review of the traditional use and phytochemistry of Leonotis leonurus. *J Ethnopharmacol*. 2015; **174:** 520–39.

Ojewole JA. Antinociceptive, anti-inflammatory and antidiabetic effects of *Leonotis leonurus* (L.) R. BR. [Lamiaceae] leaf aqueous extract in mice and rats. *Meth Find Exp Clin Pharmacol.* 2005; **27:** 257–64.

Xu D, Chen M, Ren X, Ren X, Wu Y. Leonurine ameliorates LPS-induced acute kidney injury via suppressing ROS-mediated NF-κB signaling pathway. *Fitoterapia*. 2014; **97:** 148–55.

Chapter 19

Restricted and Banned Herbals

Aconitum anthora L.

Common name: Anthora, yellow monkshood, or healing wolfs bane.

Botany: A perennial herb, growing upto 25–100 cm. Leaves are alternate and petiolate, and divided several times completely to the base or midrib. Flowers yellow, perianth persistent. The fruits are (3–) 5 follicles.

Chemical composition: The root contains a large amount of volatile salt and essential oil, while the foliage and stems contain alkaloids, 10-hydroxy-8-O-methyltalatizamine (Fig. 121), isotalatisidine, hetisinone and flavonoids.

Therapeutics: *A. anthora* has been used externally against rheumatism and deep pain.

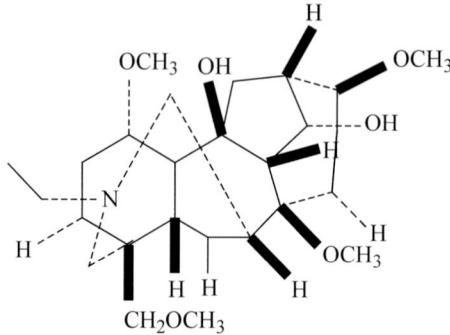

Fig. 121. Structure of 10-hydroxy-8-O-methyltalatizamine.

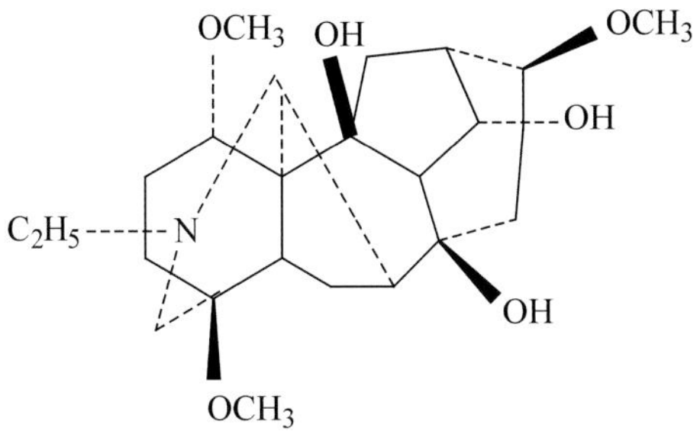

OCH₃ OH   OCH₃

---OH

C₂H₅----N

OH

OCH₃

Fig. 122. Structure of isotalatizidine.

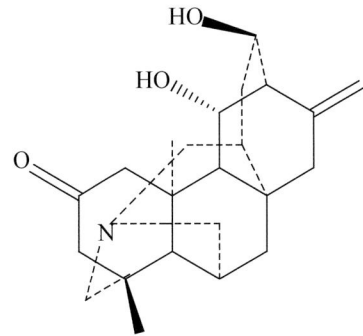

Fig. 123. Structure of hetisinone.

Aconitum balfourii Stapf

Distribution: East Asia—The Himalayas from Nepal to Tibet at an elevation of 2200–4000 metres.

Botany: An erect, glabrous shrub growing more than 1.5 m in height. Roots are fusiform and tuberous. Stem is simple or branched, about 1.2 m high, dull purplish-brown. The upper leaves have short stalk. The lower leaves have long stalk. The flowers are hermaphrodite. Inflorescence has many flowered racemes with yellowish tomentum.

Chemical composition: Tubers are rich sources of pseudoaconitine (0.4 to 0.5%) and aconite alkaloids including aconitine, benzylaconitine, picroaconitine (benzoylaconitine), and haemonepellen in traces.

Action: The root is analgesic, anti-inflammatory, antirheumatic and vermifuge.

Therapeutics: The tubers are used in Tibetan medicine where they are considered to have an acrid and sweet taste with heating potency. The root is used in the treatment of all types of pain and inflammation from gout or arthritis, all disorders due to worms or micro-organisms, amnesia, loss of bodily heat, leprosy and paralysis.

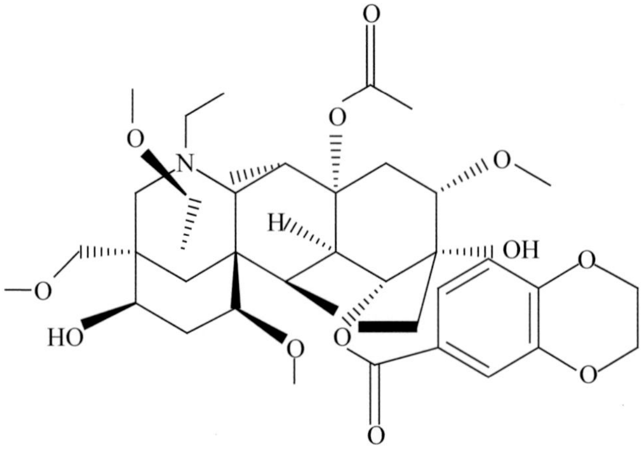

Fig. 124. Structure of pseudoaconitine.

Parts used: Tubers.

Note: *A. balfourii* is prohibited for export under the regulation of Convention of International Trade in Endangered Species (CITES).

Aconitum baikalensis Turcz.

Syn: *A. baicalense* Turcz. ex Rapaics

Common name: Baikal aconite.

Distribution: Russia.

Botany: *A. baikalensis* is a perennial herb.

Chemical composition: Alkaloids, napelline, hypaconitine, songorine, mesaconitine, 12-epinapelline N-oxide.

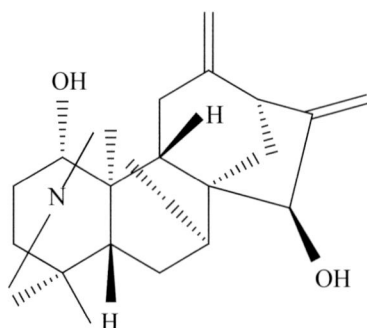

Fig. 125. Structure of songorine.

Aconitum brachypodum Diels.

Distribution: Yunnan and Sichuan Provinces of China.

Botany: The roots are carrot shaped. Stems 40–80 cm high, unbranched or branched. When wilting of flowering stems and lower leaves occurs, middle leaf has a short handle. 7-up flower racemes densely flowered. Inflorescence rachis and pedicels densely curved and appressed pubescent.

Chemical composition: Alkaloids, brachyaconitines A–D.

Actions: Anti-rheumatic and analgesic.

Therapeutics: Rheumatism and pains.

Parts used: Roots.

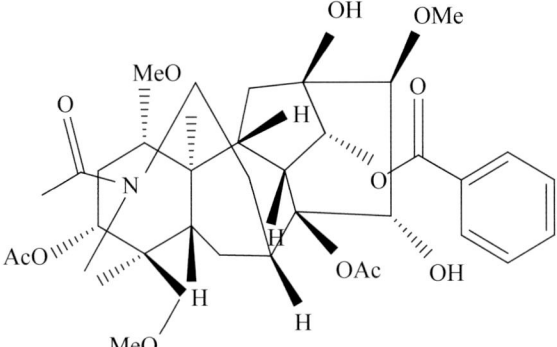

Fig. 126. Structure of brachyaconitine A.

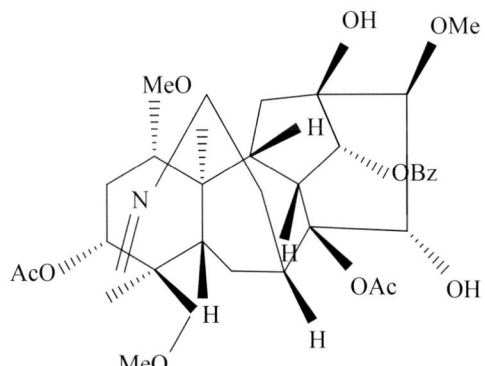

Fig. 127. Structure of brachyaconitine B.

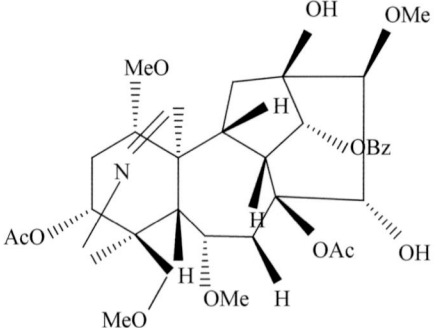

Fig. 128. Structure of brachyaconitine C.

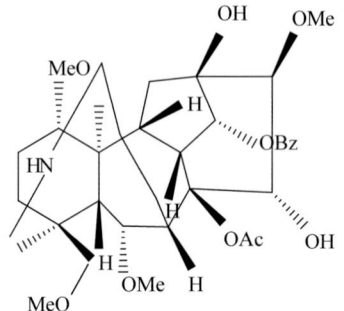

Fig. 129. Structure of brachyaconitine D.

Aconitum bulleyanum Diels.

Distribution: China.

Chemical composition: Alkaloids: yunaconitine, crassicaudine, foresaconitine, chasmaconitine, bulleyaconitine A, franchetine and beta-sitosterol.

Therapeutics: Pain associated with arthritis, influenza, rashes and snake bite.

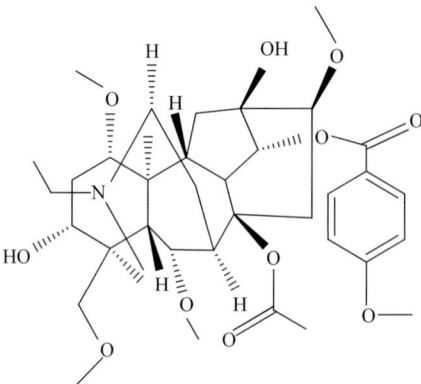

Fig. 130. Structure of yunaconitine.

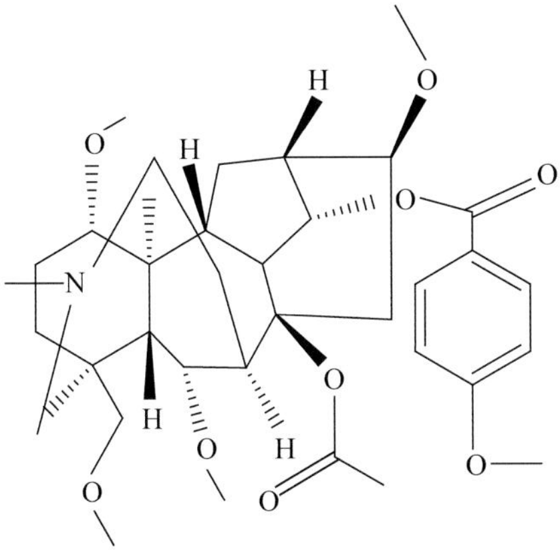

Fig. 131. Structure of foresaconitine.

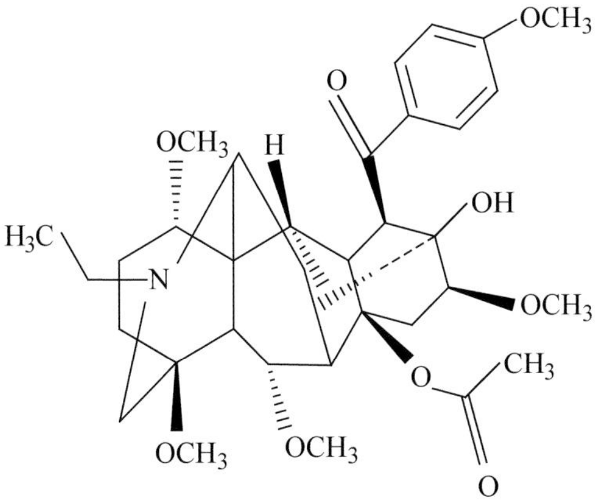

Fig. 132. Structure of bulleyaconitine A.

Aconitum carmichaelii Debeaux

Common name: Chinese aconite, Carmichael's monkshood or Chinese wolfs bane.

Distribution: Native to East Asia and eastern Russia.

Botany: *A. carmichaelii* is an erect perennial, with 3- to 5-lobed ovate, leathery leaves. Dense panicles of blue flowers are produced in late summer and autumn.

Chemical composition: Aconitine (0.004%), hypaconitine (0.12%) and mesaconitine (0.033%).

Therapeutics: *A. carmichaelii* is used topically in Dit Da Jow liniment.

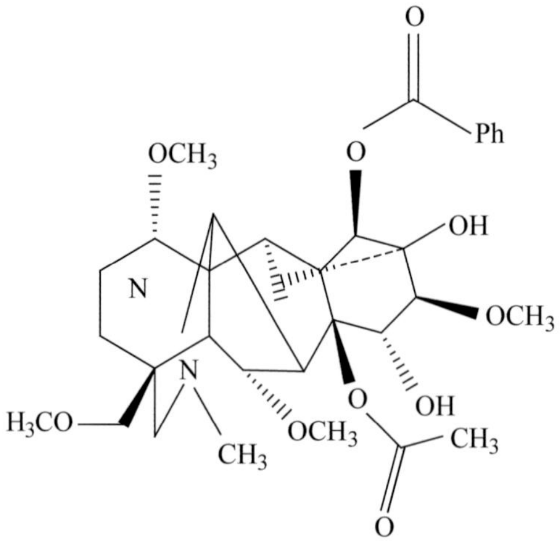

Fig. 133. Structure of hypaconitine.

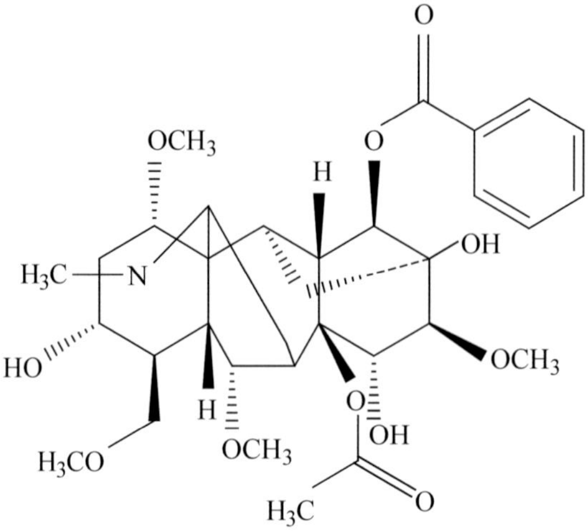

Fig. 134. Structure of mesaconitine.

Aconitum chasmanthum Stapf ex Holmes

Common name: Gaping Monkshood, Gaping Flower Aconite.

Distribution: E. Asia—Western Himalayas from Chitral to Kashmir at 2100–3500 metres.

Botany: A perennial herb growing upto 90 cm. The roots are tuberous and stem leafy. The leaves have regular pattern, deeply cut. The upper leaves are slightly smaller than lower ones. The flowers blue or purplish, borne in 30 cm long racemes.

Chemical composition: Aconitine (0.78–0.81%).

Actions: The dried root is analgesic, anodyne, diaphoretic, diuretic, irritant and sedative.

Therapeutics: The dried pulverised roots are mixed with butter and used as ointment on abscess and boils. They are also mixed with tobacco and used as "errhine".

Note: *A. chasmanthum* is prohibited for export under the regulation of Convention of International Trade in Endangered Species (CITES). It has been included in IPL, IPC and IP, and also included in Schedule E (1) of the Drug and Cosmetic Act, 1940 as a poisonous plant.

Aconitum coreanum (H. Léveillé) Rapaics

Distribution: Grassy slopes, forests; 200–900 m. N Hebei, E Heilongjiang, Jilin, Liaoning [Korea, Mongolia, Russia (Far East)].

Botany: A perennial herb growing up to 100 centimeters. The roots are fusiform. Stem simple or branched. Leaves are alternate and the upper one are shorter and sessile. The flowers are pale yellow in colour.

Actions: Anti-arrhythmic, analgesic and anti-inflammatory.

Therapeutics: Cardialgia, facial distortion, epilepsia, migraine headache, vertigo, tetanus, infantile convulsion and rheumatic arthralgia.

Aconitum deinorrhizum Stapf

Distribution: Sub-alpine and alpine zone of the Himalayas from Indus to Kumaon occurring at altitudes between 2400–4500 m.

Botany: A biennial herb. Roots are paired and stem erect. Leaves are scattered. Inflorescence simple. Follicles 3, greyish pubescent. Seeds obconic.

Actions: Anodyne, diuretic and diaphoretic.

Therapeutics: Externally, it is used with mustard oil for massage in neuralgia, paralysis and muscular rheumatism. Root is smoked during toothache and body pain. Leaves are employed to improve the flavour of the country liquor. It is also used in leprosy, cholera and in diarrhoea.

Note: *A. deinorrhizum* is in the negative list of export. It is one of the endangered medicinal plants of the Western Himalayas.

Aconitum delavayi Franch

Distribution: Hengduan mountains of China.

Therapeutics: Rheumatism, traumatic injuries, blood stasis, swelling, pain, hematemesis, hemoptysis, hematochezia, piles and haemorrhage.

Aconitum duclouxii Levl.

Distribution: North-western Yunnan.

Botany: *A. duclouxii* is a perennial herb.

Chemical composition: Alkaloids: benzoylaconine, N-deethylaconitine, aconitine, deoxyaconitine and ducloudine A.

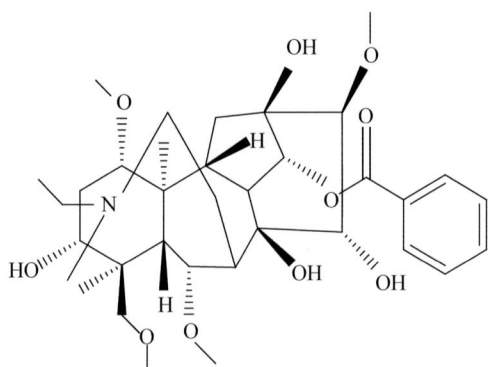

Fig. 135. Structure of benzoylaconine.

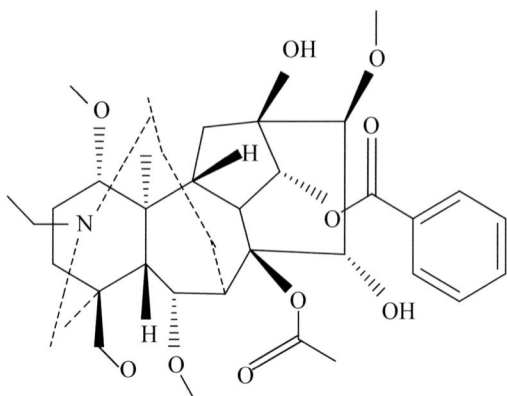

Fig. 136. Structure of deoxyaconitine.

Aconitum ferox Wall. ex Ser.

Common name: Teliya bish (Hindi), Vatsnabha (Sanskrit) and Indian aconite (English).

Distribution: *A. ferox* is common in Nepal at 3,000 m. It is also found in the Himalayas and Kashmir.

Botany: *A. ferox* is a perennial plant that grows up to one meter in height. It has tuberous roots that are dark brown on the outside and yellow on the inside. The leaves are larger towards the bottom, growing smaller and shorter towards the top of the plant. The flowers are purple-blue and located at the end of the stems. The fruit is a tube-like capsule that opens at the top.

Chemical composite: Alkaloid (napelline or pseudoaconitine).

Therapeutics: Externally, the drug is rubbed as liniment in arthritis and rheumatism. It is poisonous and rarely administered internally.

Aconitum fischeri Rchb.

Common name: American Aconite, Monkshood-Azure.

Distribution: North America.

Botany: *A. fischeri* sends up striking upright spikes of lavender-blue flowers from late summer to early fall when few other blue flowers are in bloom. Each individual flower is shaped like a little helmet or hood, thus the common name Monkshood.

Chemical composite: Aconitine and japaconitine (identical with benzoylaconine). By saponification, japaconitine is resolved into japaconine and benzoic acid.

Therapeutics: According to the ancient Chinese medical lore, it is used in the treatment of cold, cough, and fever.

Aconitum gymnandrum Max.

Distribution: China.

Chemical composition: Flavonoids.

Actions: Carminative and analgesic.

Aconitum heterophylloides (Brühl) Stapf

Syn: *A. leucanthum* (Brühl) Stapf.

Common name: Nepal Monkshood.

Distribution: The Himalayas, from Indus to Kumaon, Himachal Pradesh, at altitudes of 2400–3800 m.

Botany: *A. heterophylloides* is a tall hairless, biennial herb with paired tuberous roots. Leaves are scattered, kidney-shaped or ovate-kidney-shaped. Flowers are more than 2.5 cm long, bright blue to greenish blue. Fruit is a capsule bearing numerous seeds.

Chemical composition: The roots contain 0.9% total alkaloids, of which 0.51% are pseudoaconitine.

Therapeutics: Roots and leaves are used in treating rheumatism, rheumatic fever and acute headache.

Aconitum heterophyllum Wall. ex Royle

Common name: Indian Atis.

Botany: *A. heterophyllum* is a small plant with an erect stem but is sometimes branched. The leaves of this herbaceous plant are heteromorphous, as the lowest leaves have long petioles and the uppermost leaves are amplexicaul. The flowers are violet or blue in colour. The fruit follicles are 16–18 cm long.

Chemical composition: Alkaloids (atisine, hetisine, and heteratisine), tannins, pectin, starch and fat and mucilage.

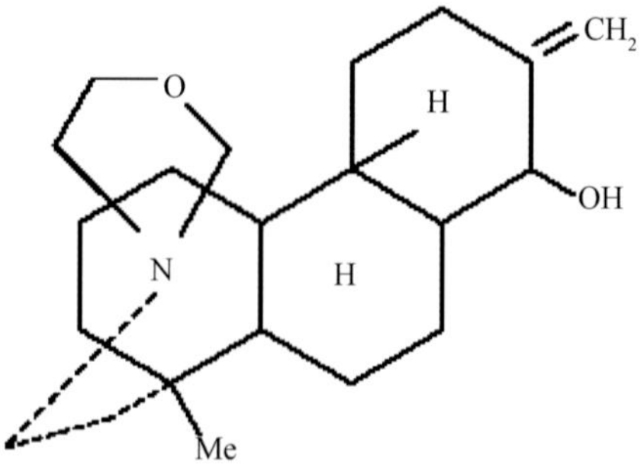

Fig. 137. Structure of atisine.

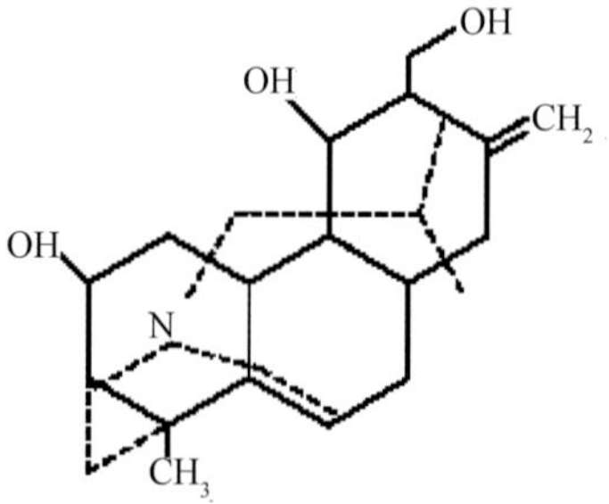

Fig. 138. Structure of hetisine.

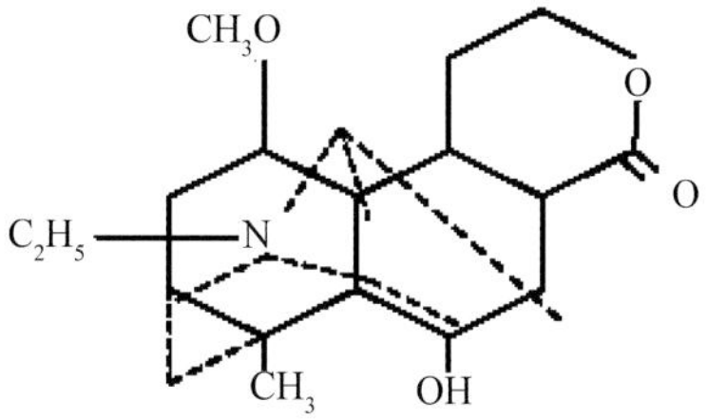

Fig. 139. Structure of heteratisine.

Action: Appetiser.

Therapeutics: Loss of appetite, convalesce, dyspepsia, malaria and childhood diseases.

Parts used: Roots.

Aconitum jaluense Komarov

Distribution: Korea, Russia (Far East).

Botany: Caudex conical, ca. 3 cm. Stem 45–100 cm, glabrous, usually branched, with leaves equally arranged along stem. Proximal cauline leaves withered at anthesis; middle ones with petiole 3–6 cm; leaf blade pentagonal, 7–12 × 8–16 cm, abaxially glabrous, adaxially rarely appressed pubescent, base cordate, 3-sect; central segment rhombic, apex acuminate, 3-fid; lateral segments obliquely flabellate, unequally parted into two halves. Inflorescence terminal or axillary, several or many flowered; rachis and pedicels usually densely spreading pubescent.

Chemical composition: Alkaloids: lipomesaconitine, napelline, lipohypaconitine and hokbusine A.

Aconitum kirinense Nakai

Distribution: China, Russia (Far East).

Botany: Rhizome unknown. Stem 80–120 cm tall, 3–5.5 mm in diam., branched, basally sparsely spreading yellow villous, apically retrorse yellow pubescent. Basal leaves ca. 2, and proximal cauline leaves long petiolate; petiole 20–30 cm, sparsely spreading pilose or nearly glabrous. Cauline leaves 2–6; leaf blade reniform-pentagonal, 12–17 × 20–24 cm, abaxially sparsely pilose at veins or nearly glabrous, adaxially appressed and retrorse pubescent, 3-parted. Inflorescence 18–22 cm, many flowered; rachis and pedicels retrorse and appressed pubescent; proximal bracts leaflike, others linear.

Chemical composition: Alkaloids: kirinines A–C.

Therapeutics: Rheumatic arthritis and rheumatoid arthritis.

Aconitum kusnezoffii Rchb

Distribution: East Asia—Northern China, Northern Japan in Kamtschatka, Korea and Siberia.

Botany: Caudex conical or carrot-shaped, 2.5–5 cm, 7–12 mm in diam. Stem (65–) 80–150 cm, usually branched, glabrous, with leaves equally arranged along the stem. Proximal cauline leaves withered at anthesis, middle ones shortly to long petiolate; petiole 3–11 cm, glabrous; leaf blade pentagonal, 9–16 × 10–20 cm, papery or subleathery, abaxially glabrous, adaxially sparsely retrorse pubescent, base cordate, 3-sect; central segment rhombic, apex acuminate, subpinnately divided or lobed; lateral segments obliquely flabellate, unequally 2-parted.

Chemical composition: Alkaloids, asconine, aconitine, mesaconitine, hypaconitine, benzoylaconine, benzoylmesaconine, benzoylhypaconine and beiwudine.

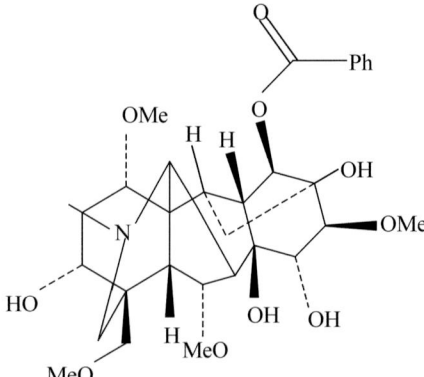

Fig. 140. Structure of benzoylmesaconine.

Actions: Analgesic and anti-rheumatic.

Therapeutics: Used to treat heart failure congestion, neuralgia, rheumatism, and gout in Homeopathic system of medicine.

Aconitum laeve Royle

Common name: Grape-Leaved Monkshood.

Distribution: The Himalayas, from Kashmir to Western. Nepal, at altitudes of 2900 m.

Botany: *A. laeve* is a perennial herb with elongated and cylindrical root. Stem is erect, up to 1.5 m tall, simple or with a few branches, the upper part is hairy and viscid and hairless below. Basal leaves have a very long stalk, usually wither at flowering time.

Stem leaves gradually decrease in size towards inflorescence, deeply 5–9-lobed with obovate to obovate-oblong segments, coarsely dentate or sometimes incised again, circular to kidney-shaped in outline, up to 30 cm wide, hairless, uppermost merging with floral leaves.

Chemical composition: Alkaloids: watinine, delphatine, lappaconitine, puberanine and N-acetylsepaconitine.

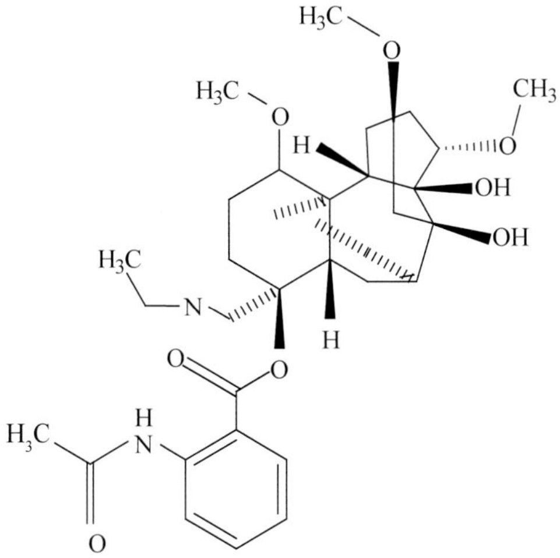

Fig. 141. Structure of lappaconitine.

Actions: Anti-inflammatory antioxidant and tyrosinase inhibition.

Aconitum lycoctonum L.

Common name: Northern wolfs bane. Hindi: Bikh.

Distribution: Native to Europe and northern Asia.

Botany: *A. lycoctonum* is an herbaceous perennial plant growing up to 1 m tall. The leaves are palmately lobed with four to six deeply cut lobes. The flowers are 18–25 mm long, dark violet and rarely pale yellow.

Chemical composition: Lycoctonine, 6-O-acetyldemethylenedelcorine, 6-O-acetyl-14-O-methyldelphinifoline, 14-O-methyldelphinifoline and gigactonine.

Actions: The root is alterative, anaesthetic, antiarthritic, antitussive, deobstruent, diaphoretic, diuretic, sedative and stimulant.

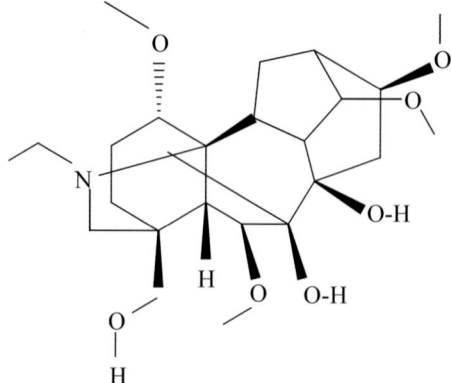

Fig. 142. Structure of lycoctonine.

Aconitum nagarum Stapf

Distribution: Manipur (Indo–Himalaya).

Chemical composition: Aconitine.

Actions: Antibacterial.

Aconitum napellus L.

Common name: Blue Rocket, Friar's Cap, Auld Wife's Huid.

Distribution: Native and endemic to western and central Europe.

Botany: *A. napellus* is a herbaceous perennial plant growing up to 1 metre tall, with hairless stems and leaves. The leaves are rounded, 5–10 cm diameter, palmately divided into five to seven deeply lobed segments. The flowers are dark purple to bluish-purple, narrow oblong helmet-shaped, 1–2 cm tall.

Chemical composition: Alkaloid: napelline.

Actions: Anodyne, diuretic and diaphoretic.

Aconitum nasutum Fisch. ex Rchb.

Distribution: Caucasus (northern and eastern regions).

Botany: Roots rounded, tuberous. Stem erect, branched, grows up to 1 m, glabrous. Leaves palmate, lobed, the lobes linear-lanceolate and acute. Inflorescence a long, loose raceme. Flowers numerous, large, 3–4 cm long, light blue or violet.

Chemical composition: Alkaloids: aconasutine, trabzonine, lappaconitine, lycoctonine, gigactonine, pseudokobusine and septatisine.

Aconitum naviculare (Brühl) Stapf

Distribution: Trans-Himalayan region of Nepal.

Chemical composition: Alkaloid: navirine and flavonoid glycosides.

Aconitum orochryseum Stapf

Distribution: Bhutan.

Chemical composition: Alkaloids: orochrine, 2-O-acetylorochrine, and 2-O-acetyl-7α-hydroxyorochrine, atisinium chloride and virescenine.

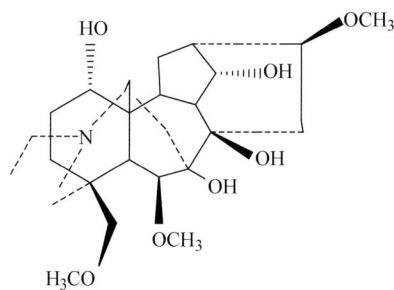

Fig. 143. Structure of virescenine.

Therapeutics: Common cough and cold, bilious fever, dysentery, as an antidote for snake bite and also as a febrifuge for fever associated with malaria infection, kidney dysfunction and stomach ulceration.

Aconitum gammiei

Distribution: Nepal.

Chemical composition: Alkaloids: 16-Acetoxycardiopetaline, 15-Acetyl-13-dehydrocardiopetamine, 15-Acetylcardiopetamine, brachyaconitine, cardiopetamine, *N*-Deethylaconitine, 1, 14-Diacetylneoline, 12-Epiacetyldehydronapelline, ephedrine, hokbusine A, senbusine A, lpaconitine, merckonine, myriophyllosides F and napelline.

Aconitum ouvrardianum Hand.-Mazz

Distribution: Yunnan.

Chemical composition: Alkaloids: ouvrardiantine and ouvrardiandines A and B.

Aconitum palmatum D. Don

Synonym: *Aconitum bisma* (Buch.-Ham.) Rapaics.

Common name: Crowfoot (English), Prativisha (Sanskrit).

Distribution: Eastern Asia—The Himalayas in Nepal, Sikkim and southern Tibet. Alpine regions between 3,000 and 5,000 metres.

Botany: *A. palmatum* is biennial & perennial herb with tuberous and paired roots. The leaves of the orbicular–cordate to reniform. Flowers are greenish blue in color and hermaphrodite. The follicles are 2.5–3.0 cm long. The seeds are blackish.

Chemical composition: Alkaloids: vakognavine, palmasine, vakatisine, vakatisinine and vakatidine.

Therapeutics: The root powder of *A. palmatum* is mixed with water and the diction is taken as an antidote in food poisoning and snake bite in Makalu-Barun and Kangchenjunga in Nepal. The root powder is also used to treat fever, headache and stomach ache.

Aconitum piepunense Hand-Mazz

Distribution: Yunnan Province.

Botany: A perennial herb.

Chemical composition: Alkaloids: piepunensine A, and 18-acetylcammaconine, pengshenine B, talatisamine, aconosine, yunaconitine, and talatizidine.

Part used: Roots.

Aconitum soongaricum Stapf

Distribution: Central-Asiatic provinces of the USSR (Tien-Shan, Dzungaria) and Turkestan.

Botany: A perennial herb.

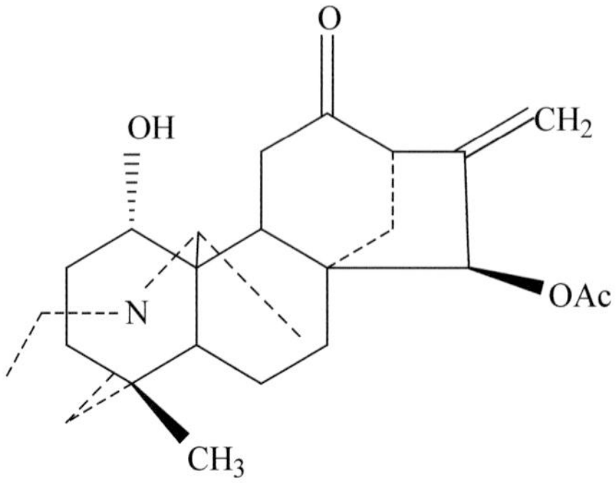

Fig. 144. Structure of acetylsongorine.

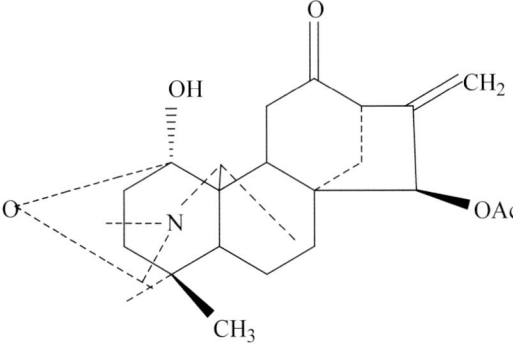

Fig. 145. Structure of songoramine.

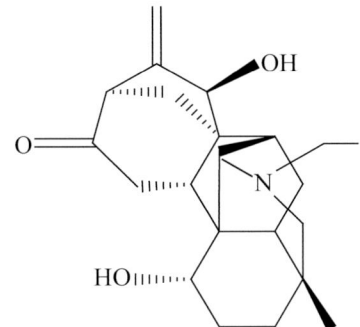

Fig. 146. Structure of songorine.

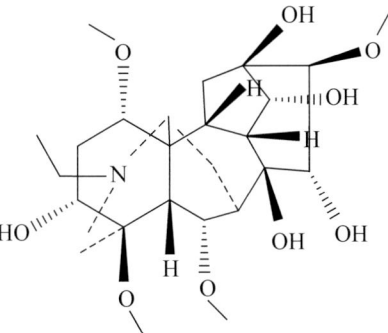

Fig. 147. Structure of aconine.

Chemical composition: Alkaloids: Acetylsongorine, songoramine, songorine, songorinine, 12-Acety-12-epinapelline and aconine.

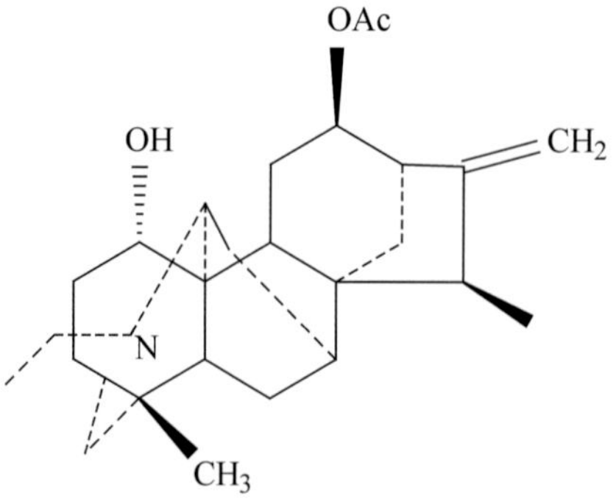

Fig. 148. Structure of 12-Acetyl-12-epinapelline.

Aconitum sungpanense var. *leucanthum* W.T. Wang

Distribution: The south-western part of China and especially abundant in Sichuan province.

Botany: A perennial herb.

Chemical composition: Alkaloids, leucanthumsines A–E.

Action: Analgesic.

Therapeutics: Rheumatism, arthritis, and neurological disorders.

Aconitum taipaicum Hand.-Mazz

Distribution: Southern Shaanxi yield (Qinling), Western Henan (Luanchuan). Health elevation of 2600–3400 meters mountain meadow.

Botany: A perennial herb.

Chemical composition: Alkaloids: atisine, delfissinol, liangshanine, hypaconitine, isodelelatine and delelatine.

Actions: Anti-inflammatory and analgesic.

Aconitum transsectum Diels

Distribution: North-western Yunnan.

Botany: A perennial herb.

Chemical composition: Alkaloids, aconitramines D–E.

Therapeutics: Embolism and rheumatism.

Part used: Roots.

Aconitum vilmorinianum Komarov

Distribution: Yunnan Province, China.

Chemical composition: Alkaloids including vilmorrianines E–G, and a new natural alkaloid N-desethyl-N-formyl-8-O-methyltalatisamine.

Therapeutics: Rheumatism and pain.

Aconitum violaceum Jacq.

Common name: Patis, Violet Monkshood.

Distribution: The Himalayan region of India, Pakistan and Nepal. Within India, it has been recorded in the alpine slopes in an altitude range of 3600–4800 m.

Botany: *A. violaceum* is a very variable perennial herb. Stem is 10–30 cm tall. Leaves have a round blade and are palmately cut at the base. Flowers are 2–2.5 cm.

Chemical composition: Indaconitine.

Action: Bitter, cooling, anti-inflammatory and febrifuge.

Therapeutics: *A. violaceum* is used in the treatment of snake and scorpion bites, contagious infections and inflammation of the intestines.

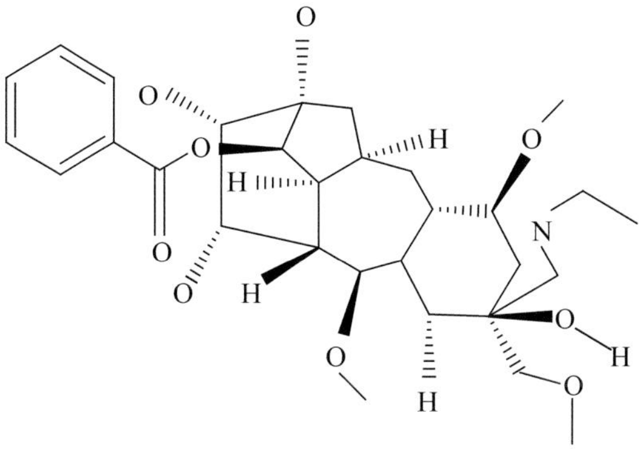

Fig. 149. Structure of indaconitine.

Aconitum vulparia Rchb

Common name: Wolf's bane.

Botany: *A. vulparia* is a hardy perennial plant. Long-lived clumps produce very dense sprays of pale sulphur yellow hooded flowers in spring and early summer, attractive bright green shiny foliage.

Chemical composition: Alkaloids, aconorine, lexhumboldtine, lappaconitine, anthranoyllycoctonine, lycoctonine, puberaconitine, ajacine, and septentriodine.

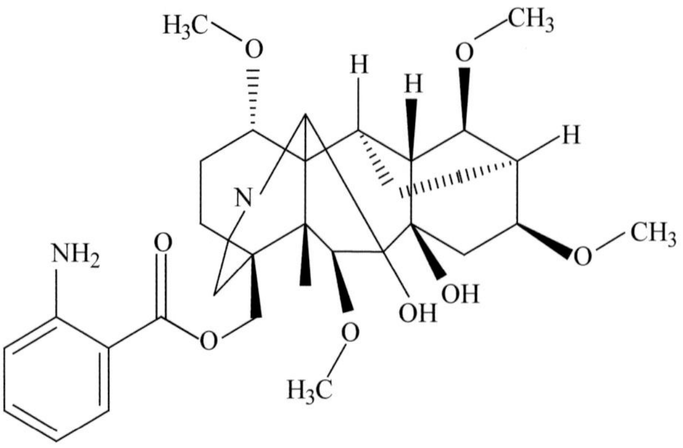

Fig. 150. Structure of anthranoyllycoctonine.

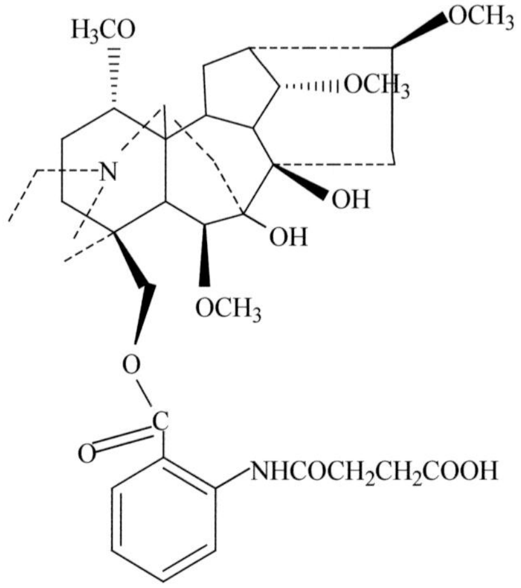

Fig. 151. Structure of puberaconitine.

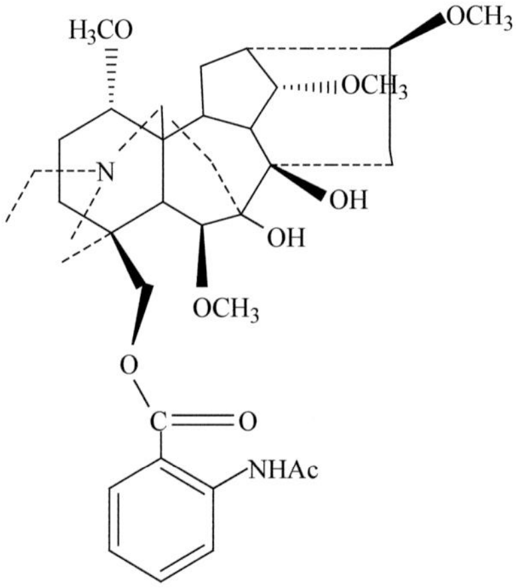

Fig. 152. Structure of ajacine.

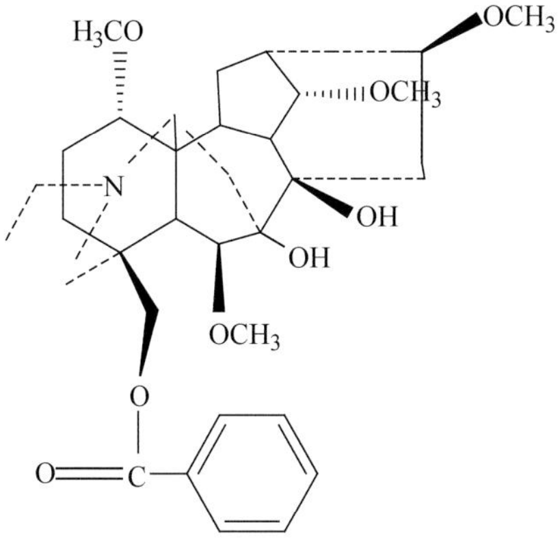

Fig. 153. Structure of septentriodine.

Therapeutics: Rheumatism, neuralgia and chronic skin disorders.

Adonis vernalis L.

Syn: *Adonanthe vernalis* (L.) Spach, *Chrysocyathus vernalis* (L.) Holub.

Family: Ranunculaceae.

Common name: Pheasant's eye, spring pheasant's eye, yellow pheasant's eye and false hellebore.

Distribution: Pannonic area of Central Europe, Eastern Europe and West Siberia.

Botany: A perennial herb, 10–40 cm high, with an erect stem. The leaves are sessile, finely divided, flowers solitary to 8 cm in diameter, of 10–20 elongated yellow petals. The fruit is an achene.

Chemical composition: Cardiac glycosides: adonidin.

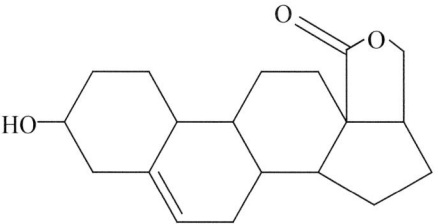

Fig. 154. Structure of adonidin.

Actions: Cardiotonic, diuretic and stimulant.

Therapeutics: In heart ailments.

Parts used: Flowers.

Note: *A. vernalis* is a strictly protected plant in some countries (CZ, SK).

Akebia quinata (Houtt.) Decne.

Syn: *Rajania quinata* Houtt.

Family: Lardizabalaceae.

Common name: Chocolate vine, Five-leaf akebia, Japanese honeysuckle.

Habitat: Native to China, Japan and Korea and naturalised in Eastern America.

Botany: A shrub growing up to 30 cm.

Description: *A. quinata* grows to 10 m (30 ft) or more in height and has compound leaves with five leaflets. The flowers are clustered in racemes and are chocolate-scented, with three or four sepals. The fruits are sausage-shaped pods which contain edible pulp.

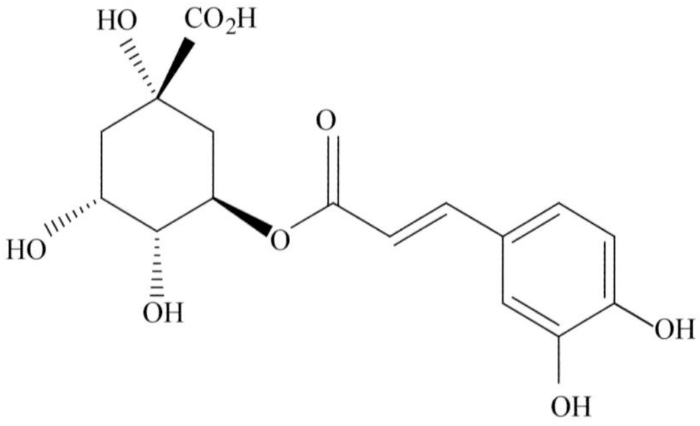

Fig. 155. Structure of 3-*O*-caffeoylquinic acid.

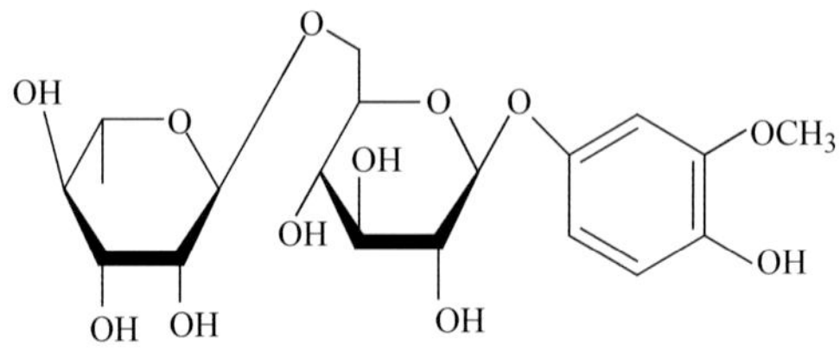

Fig. 156. Structure of cuneataside D.

Chemical composition: Lignan glycosides, akeqintoside A–C, megastigmane glycoside: akequintoside D, roseoside II, 3-*O*-caffeoylquinic acid (Fig. 155), methyl-3-O-caffeoylquinate, 3,4,5-trimethoxyphenyl-β-D-glucopyranoside, cuneataside D (Fig. 156), and 3,4-dimethoxyphenyl-6-O-(α-L-rhamnopyranosyl)-β-D-glucopyranoside (Jin et al. 2014).

Actions: As per Chinese herbology, *A. quinata* is analgesic, anti-inflammatory, diuretic, and galactatgouge.

Therapeutics: Failing lactation.

Parts used: Woody stem.

Preparation: Decoction.

Pharmacology: Akeqintosides have IL-6 inhibitory activity (Jin et al. 2014).

Legal status: *A. quinata* is prohibited in all unlicensed medicines.

Akebia trifoliata (Thunb.) Koidz.

Syn: *Rajania quinata* Houtt.

Family: Lardizabalaceae.

Common name: Three-leaf akebia.

Habitat: Native to China, Japan and Korea and naturalised in Eastern America.

Botany: A rare chocolate coloured vine. The leaves have 3–5 leaflets having prominent crenations along the corner.

Description: A rare chocolate-vine hybrid similar to *A. quinata* but shows leaves with 3–5 leaflets that have prominent crenations along the margins. Akebia quinata has leaves with (3–)5(–7) leaflets that have entirely undulate margins (rarely with 1 or 2 obscure teeth).

Chemical composition: 30-noroleanane triterpenes: 2α,3β,20α-trihydroxy-30-norolean-12-en-28-oic acid (Fig. 157), 2α,3β-dihydroxy-23-oxo-30-norolean-12,20(29)-dien-28-oic acid (Fig. 158), 3β-akebonoic acid (Fig. 159), 2α,3β-dihydroxy-30-noroleana-12,20(29)-dien-28-oic acid (Fig. 160), 3α-akebonoic acid (Fig. 161) and quinatic acid (Fig. 162) (Wang et al. 2014). Oleanane triterpenoids: 2α,3β,29-trihydroxyolean-12-en-28-oic acid, 2α,3β-dihydroxy-23-oxo-olean-12-en-28-oic acid, 2α,3β,21β,22α-tetrahydroxyolean-12-en-28,29-dioic acid, maslinic acid (Fig. 163), arjunolic acid (Fig. 164), oleanolic acid, 3-epi-oleanolic acid (Fig. 165), stachlic acid A, serratagenic acid (Fig. 166), gypsogenic acid (Fig. 167), 2α,3β-dihydroxyol-ean-

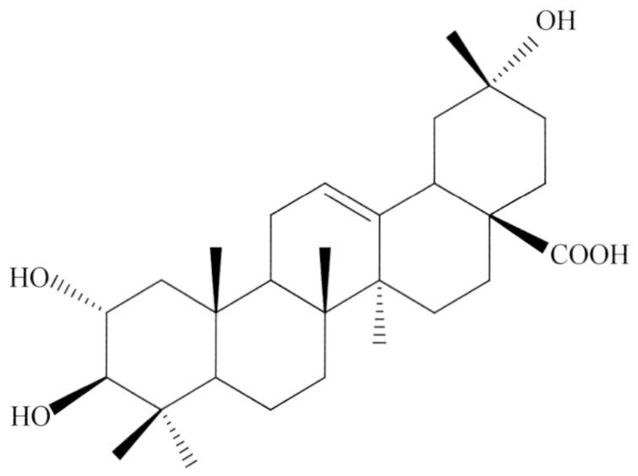

Fig. 157. Structure of 2α,3β,20α-trihydroxy-30-norolean-12-en-28-oic acid.

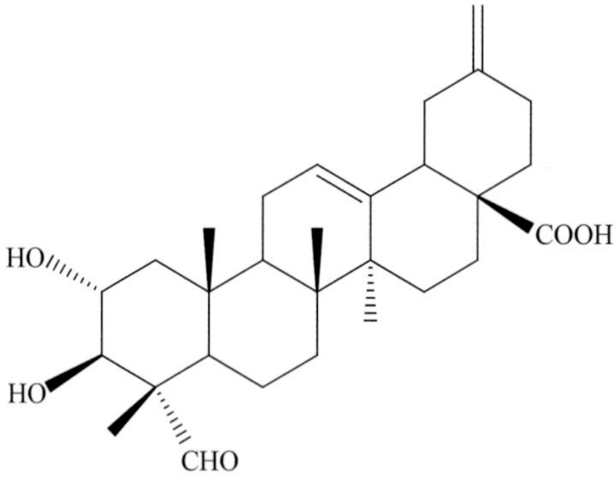

Fig. 158. Structure of 2α,3β-dihydroxy-23-oxo-30-norolean-12,20(29)-dien-28-oic acid.

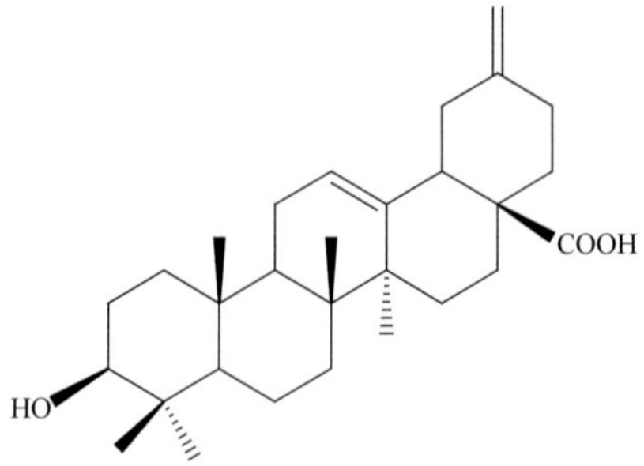

Fig. 159. Structure of 3β-akebonoic acid.

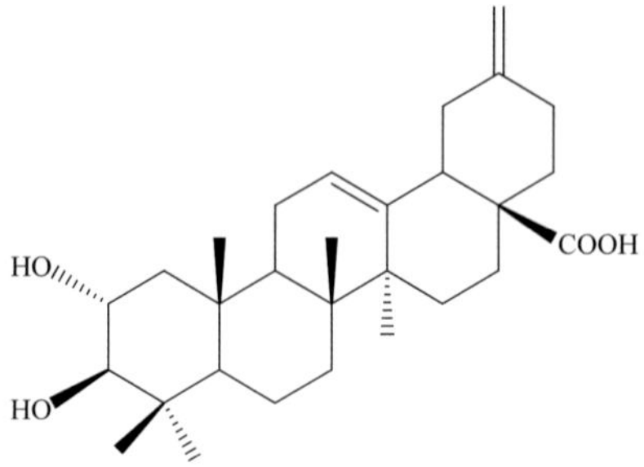

Fig. 160. Structure of 2α,3β-dihydroxy-30-noroleana-12,20(29)-dien-28-oic acid.

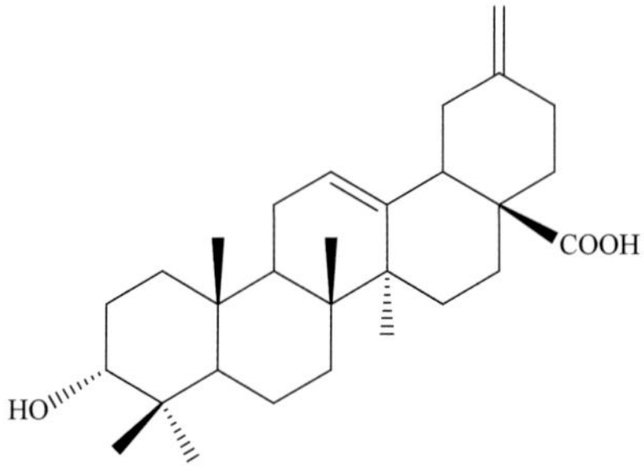

Fig. 161. Structure of 3α-akebonoic acid.

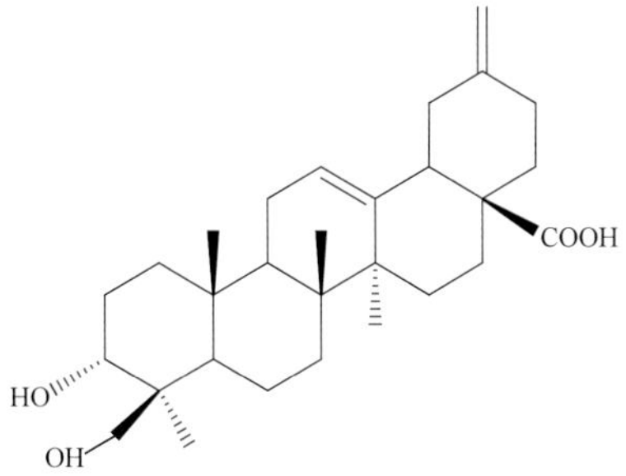

Fig. 162. Structure of quinatic acid.

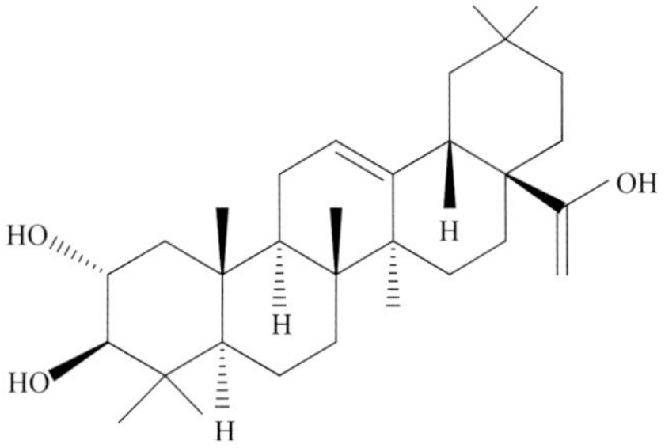

Fig. 163. Structure of maslinic acid.

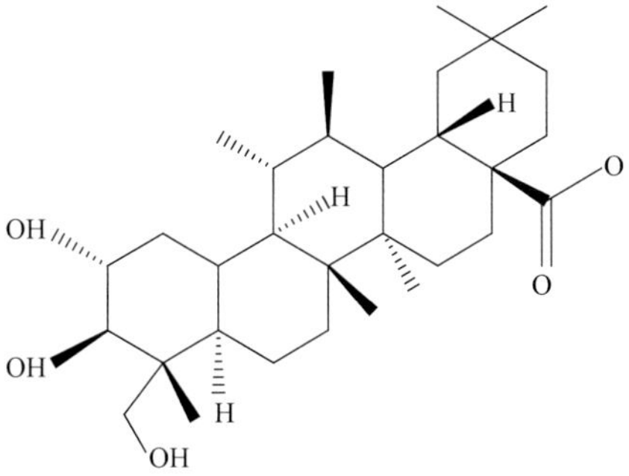

Fig. 164. Structure of arjunolic acid.

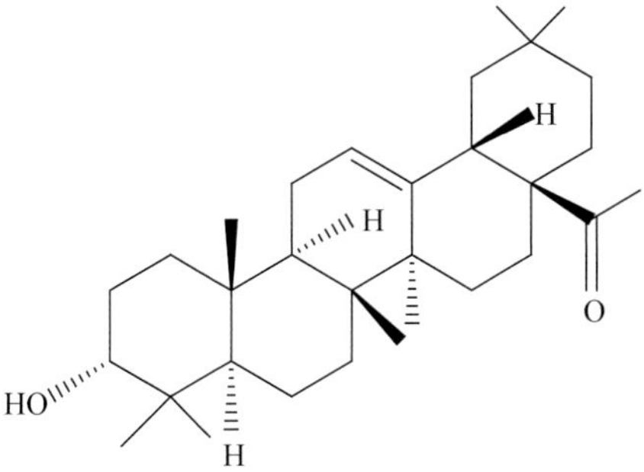

Fig. 165. Structure of 3-epi-oleanolic acid.

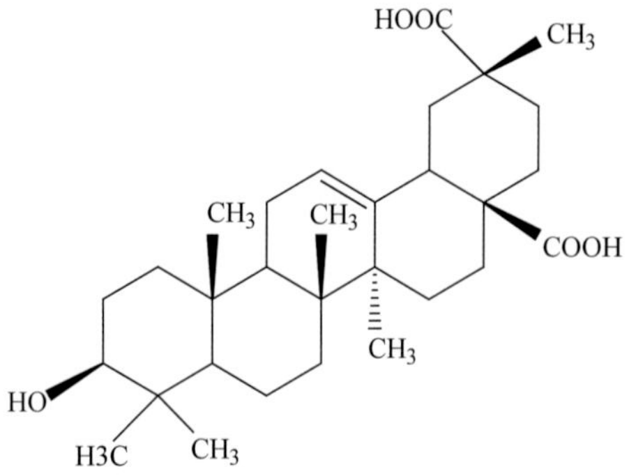

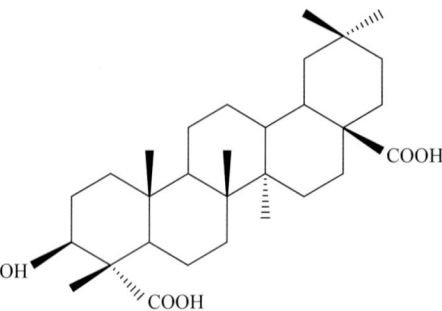

Fig. 166. Structure of serratagenic acid.

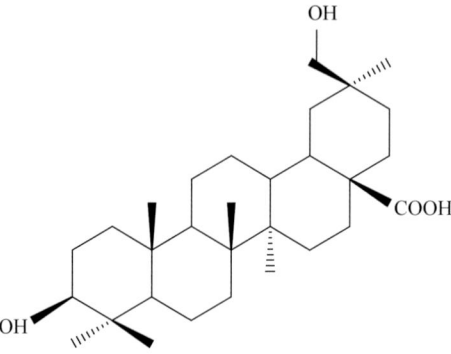

Fig. 167. Structure of gypsogenic acid.

Fig. 168. Structure of mesembryanthemoidigenic acid (3β,29-dihydroxy-olean-12-en-28-oic acid).

13(18)-en-28-oic acid, mesembryanthemoidigenic acid (Fig. 168) and 12α-hydroxy-δ-lactone (Wang et al. 2015).

Actions: Analgesic, antibacterial, antifungal, anti-inflammatory, diuretic, emmenagogue and galactogogue.

Therapeutics: Failing lactation.

Parts used: Fruit and stem.

Preparation: Decoction.

Pharmacology: 3β-akebonoic acid, 2α,3β-dihydroxy-30-noroleana-12,20(29)-dien-28-oic acid, 3α-akebonoic acid and quinatic acid showed *in vitro* bacteriostatic activity against four assayed Gram-positive bacterial strains (Wang et al. 2014). Maslinic acid, oleanolic acid and 2α,3β-dihydroxyol-ean-13(18)-en-28-oic acid have significant antibacterial activity toward all the assayed microorganisms with MIC values ranging from 0.9 to 15.6 µg/mL (Wang et al. 2015).

Legal status: *A. trifoliata* is prohibited in all unlicensed medicines.

Apocynum cannabinum L.

Common name: Amy Root, Canadian Hemp, Dogbane, Hemp Dogbane, Indian Hemp Prairie Dogbane, Rheumatism Root, and Wild Cotton.

Family: Apocynaceae.

Distribution: Throughout America.

Botany: *A. cannabinum* is a shrub growing up to 2 meters/6 feet tall. The stems are red in colour and contains a milky latex. The leaves are opposite and lanceolate. The flowers are hermaphrodite.

Chemical composition: Cardiac glycosides: cymarin, apocannoside, cyanocannoside, cannogenin and apobioside.

Actions: Hypnotic, sedative, cardiac-tonic, diaphoteric and emetic.

Therapeutics: Cough, bronchial asthma, failing lactation, rheumatism and worm-infestation.

Part used: Entire plant and roots.

Toxicity: All the parts of *A. cannabinum* are poisonous and can result in heart-arrest, if consumed internally.

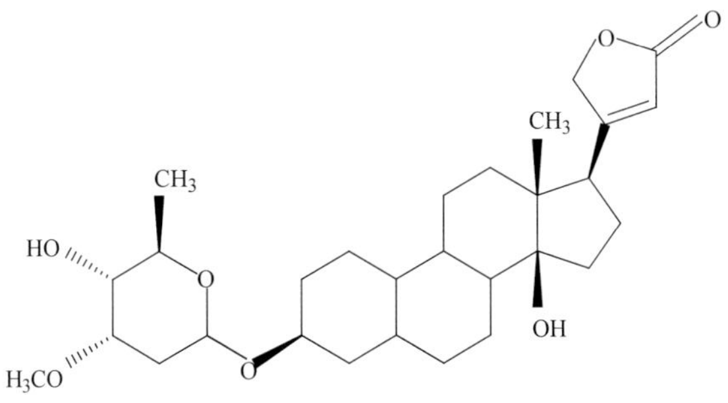

Fig. 169. Structure of cymarin.

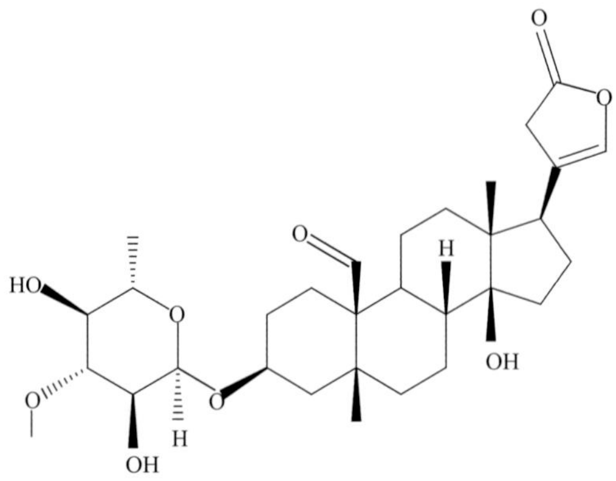

Fig. 170. Structure of cannogenin (peruvoside).

Areca catechu L.

Common name: Areca palm, areca nut palm, betel palm, Indian nut, and Pinang palm.

Family: Arecaceae.

Distribution: Tropical Pacific, Asia, and parts of east Africa.

Botany: *A. catechu* is a medium-sized palm tree, growing straight to 20 m tall, with a trunk 10–15 cm in diameter. The leaves are 1.5–2 m long, pinnate, with numerous, crowded leaflets.

Chemical composition: The major stimulant alkaloid of *A. catechu* is arecoline (up to 0.2%), the remainder of the alkaloid content (total about 0.45%) being composed of arecaidine, guvacine, and guvacoline. The seeds contain procyanidins known as arecatannins, which have been linked to carcinoma.

Actions: Astringent and myotic (constricts the pupil). Arecoline is antihelmintic and resembles in action with pilocarpine.

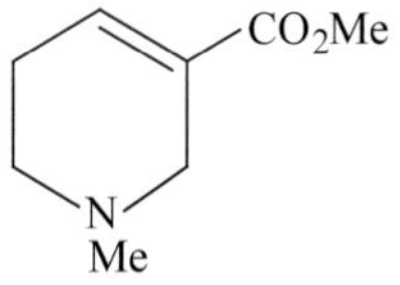

Fig. 171. Structure of arecoline.

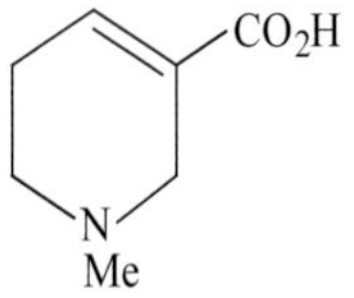

Fig. 172. Structure of arecaidine.

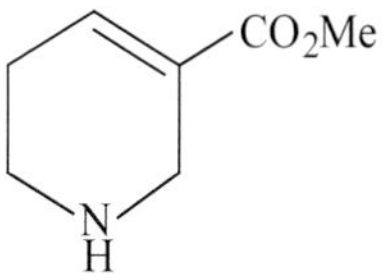

Fig. 173. Structure of guvacoline.

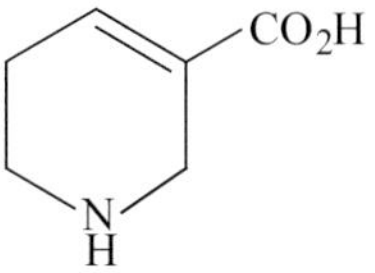

Fig. 174. Structure of guvacine.

Therapeutics: Worm infestation.

Part used: Kernels.

Aristolochia bracteata Retz.

Syn: *Aristolochia bracteolata* Lam.

Common name: Bracteated birthwort. *Kitmari* or *Dhumapatra* (Ayurveda).

Family: Aristolochiaceae.

Distribution: Distributed throughout India.

Botany: A shrub. Leaves are kidney shaped or heart shaped or rounded. Fruits are oblong capsules and with compressed seeds.

Chemical composition: Aristolochic acid I, aristolic and p-coumaric acids.

Actions: Gastric stimulant and purgative.

Therapeutics: It is used for the treatment of skin diseases, cancer, inflammation and snake-bites. Root powder is combined with honey and given internally in the case of gonorrhea, boils, and ulcers.

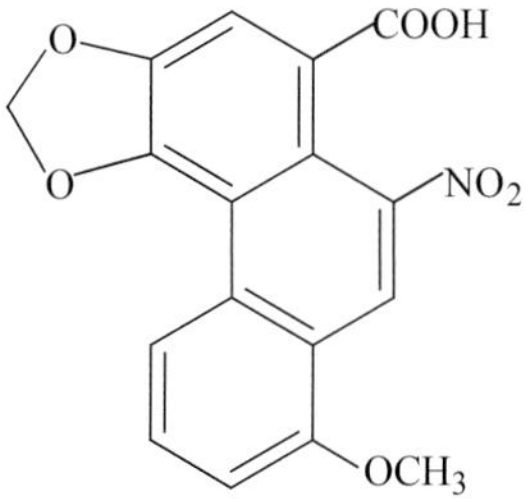

Fig. 175. Structure of aristolochic acid I.

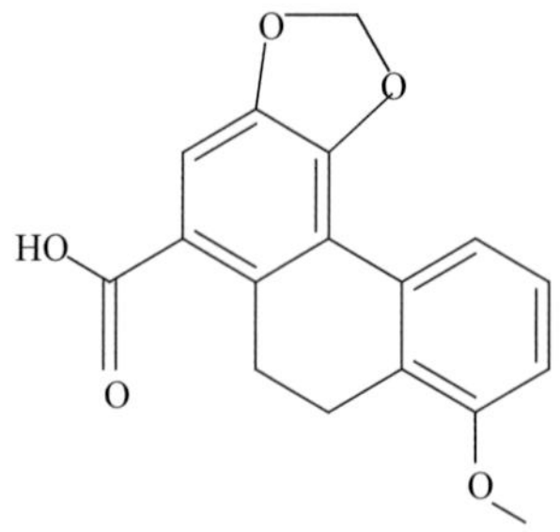

Fig. 176. Structure of aristolic acid.

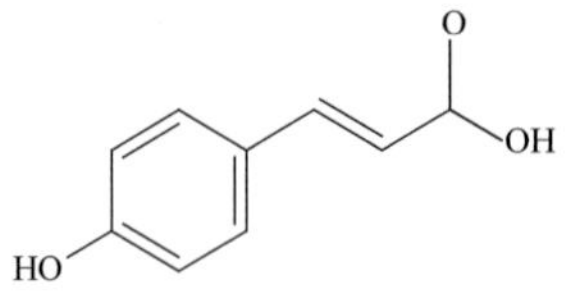

Fig. 177. Structure of p-coumaric acid.

Aristolochia clematitis L.

Family: Aristolochiaceae.

Chinese name: The European birthwort.

Distribution: Native to Europe.

Botany: A twining herbaceous plant. The leaves are heart-shaped and the flowers are pale yellow and tubular in form.

Chemical composition: Aristolochic acid.

Therapeutics: Hypertension and haemorrhoids.

Part used: Fruit.

Aristolochia contorta Bunge

Family: Aristolochiaceae.

Chinese name: Hangul.

Distribution: The mountains, fields or forests in Korea, Japan, and eastern China.

Botany: A perennial plant.

Chemical composition: Aristolochic acid and aristololactam.

Therapeutics: Hypertension and haemorrhoids.

Part used: Fruit.

Aristolochia debilis Siebold and Zuccarini

Family: Aristolochiaceae.

Chinese name: Hangul.

Distribution: Japan and China.

Botany: A perennial plant.

Chemical composition: Aristolochic acid I, aristolochic acid II, aristolochic acid III, aristolochic acid III a, aristolochic acid VII a, aristolactam I, aristolactam II and aristolactam III a.

Part used: Fruit.

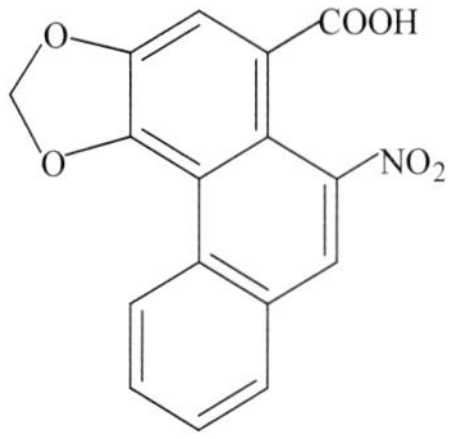

Fig. 178. Structure of aristolochic acid II.

Aristolochia fangchi Y.C. Wu ex L.D. Chou et S.M. Hwang

Family: Aristolochiaceae.

Chinese name: Guan Fang Chi.

Distribution: Dense forests or thickets, on mountain slopes, at elevations of 500–1000 metres.

Botany: A perennial climbing vine.

Chemical composition: Aristolochic acid I, aristololactam, allantoin and magnoflorine.

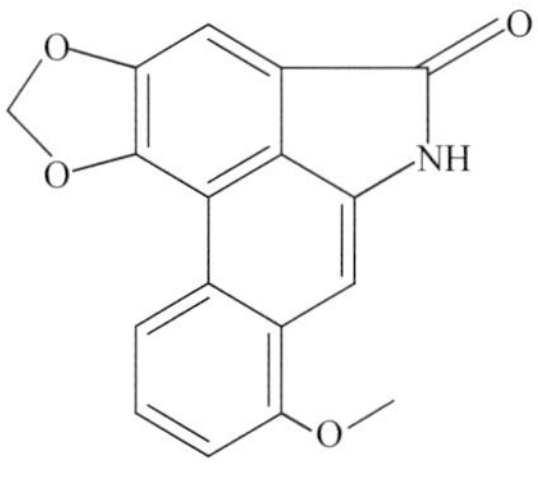

Fig. 179. Structure of aristololactam.

Therapeutics: Oedema, numbness and swelling.

Part used: Root.

Note: Because of containing aristolochic acid, *Aristolochia fangchi* herb was no longer used in pharmaceutical production and ever since *Stephania tetrandra* has become its main substitute. The notice was issued by China's State Food and Drug Administration on September 30, 2004.

Aristolochia indica L.

Family: Aristolochiaceae.

Common name: Indian birthwort. *Jata* (Ayurveda).

Distribution: Kerala in India and Sri Lanka.

Botany: A slender perennial twiner. Leaves linear-oblong to obovate-oblong abruptly or gradually obtusely acuminate. Flowers in axillary racemes, and greenish. Capsules 3.4–5 cm long, oblong or globose-oblong.

Chemical composition: Aristolochic acid, isoaristolochic acid, allantion, alkaloid, aristolochine; sesquiterpenes (ishwarone, ishwarane, aristolochene and ishwarol), tannins and an essential oil.

Actions: Stimulant, tonic and emmenagouge.

Therapeutics: It is used in traditional medicine for treating postpartum infections and snakebite. Root and leaves are used as medicine in a number of diseases such as fever, dry cough, cholera, ulcers and leprosy.

Parts used: Roots.

Aristolochia longa Linn.

Family: Aristolochiaceae.

Common name: Birthwort.

Distribution: South Europe.

Botany: The root is spindle-shaped from 5 cm to 3 dm in length, about 2 cm in thickness, fleshy, very brittle, greyish externally, brownish-yellow inside, bitter and of a strong disagreeable odour when fresh.

Chemical composition: Aristolochic acid.

Actions: Stimulant.

Therapeutics: Rheumatism.

Part used: Fruit.

Aristolochia manshuriensis Kom

Family: Aristolochiaceae.

Common name: Manchurian birthwort.

Distribution: North Korea.

Botany: A hardy climber with heart-shaped apple green leaves and large flowers.

Chemical composition: Aristolochic acid.

Part used: Fruit.

Aristolochia rotunda Linn.

Family: Aristolochiaceae.

Common name: Round-leaved birthwort.

Distribution: South Europe.

Botany: A perennial herb.

Chemical composition: Aristolochic acid.

Part used: Fruit.

Aristolochia serpentaria L.

Family: Aristolochiaceae.

Common name: Snakeroot.

Distribution: The Central and Southern United States.

Botany: The rhizome is short, horizontal rhizome having several roots below. The flowers are small and brownish-purple.

Chemical composition: Aristolochic acid and arirstolochine.

Actions: Stimulant, tonic and diaphoretic.

Therapeutics: Intermittent fever.

Part used: Dried rhizomes.

Artemisia cina Berg and C.F. Schmidt ex Poljakov

Family: Asteraceae.

Common name: Santonica.

Distribution: Native to China, Kazakhstan, and Kyrgyzstan.

Botany: *A. cina* is a small perennial semi-shrub. The stems bear on many branches and small bi- to multi-pinnatifid leaves. The flower heads are small.

Chemical composition: Santonin.

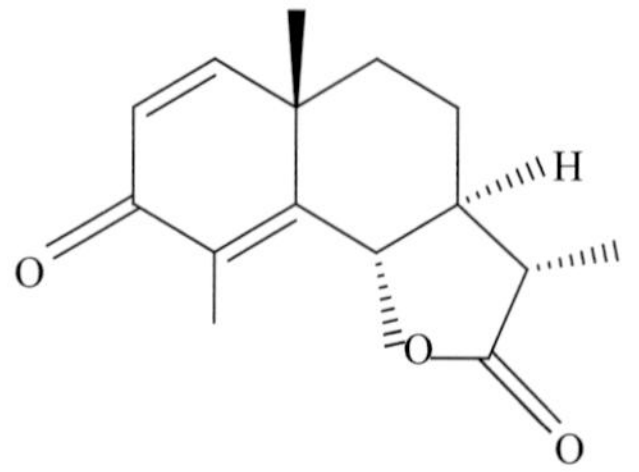

Fig. 180. Structure of santonin.

Actions: Vermifuge and febrifuge.

Therapeutics: Roundworm and thread worm infestation.

Part used: Flowerhead and seeds.

Aspidosperma quebrachoblanco Schltr.

Syn: *Macaglia quebrachoblanco* (Schltdl.) A.Lyons, *Aspidosperma quebracho* Griseb, *Macaglia quebracho* (Griseb.) Kuntze and *Aspidosperma crotalorum* Speg.

Family: Apocynaceae.

Common name: White quebracho.

Distribution: Native to Brazil, N Argentina, Bolivia, Paraguay and Uruguay.

Botany: An evergreen tree which sometimes rises to 100 feet, with an erect stem and wide-spreading crown.

Chemical composition: Cyclitol: quebrachitol and alkaloids: aspidospermine, aspidospermatine, aspidosamine, quebrachine, hypoquebrachine and quebrachamine.

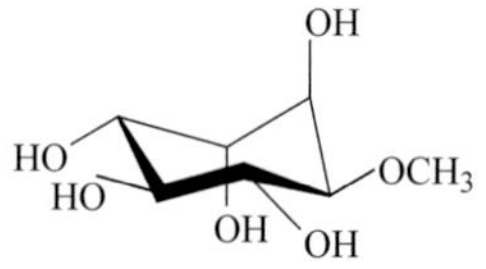

Fig. 181. Structure of quebrachitol.

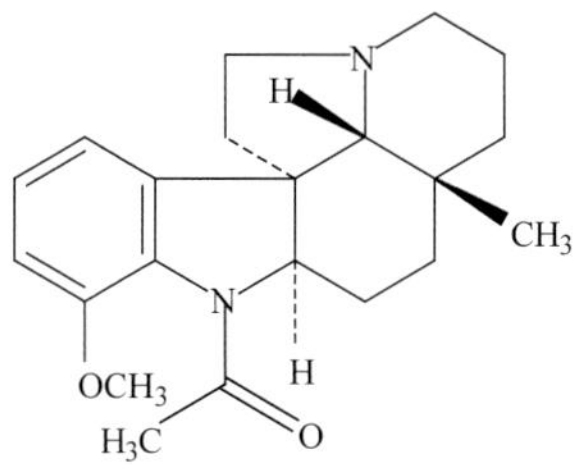

Fig. 182. Structure of aspidospermine.

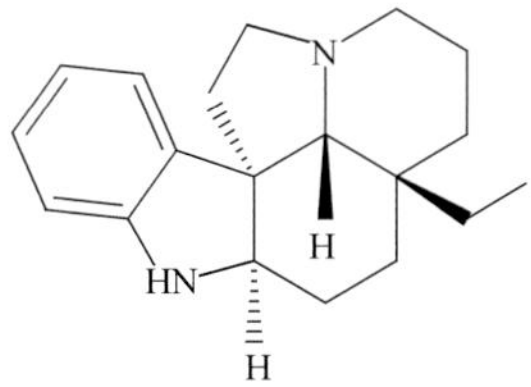

Fig. 183. Structure of aspidospermatine.

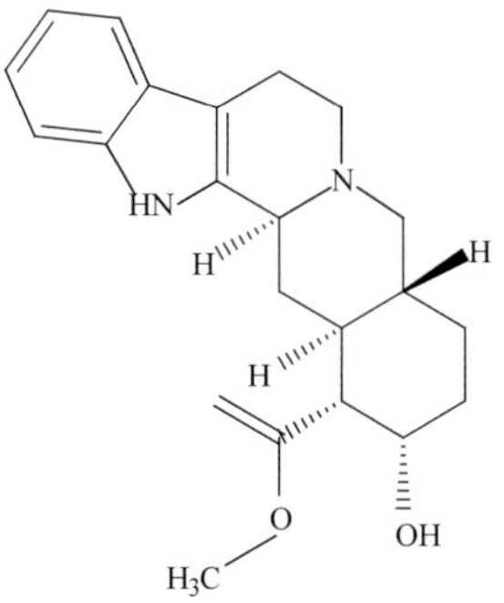

Fig. 184. Structure of quebrachine.

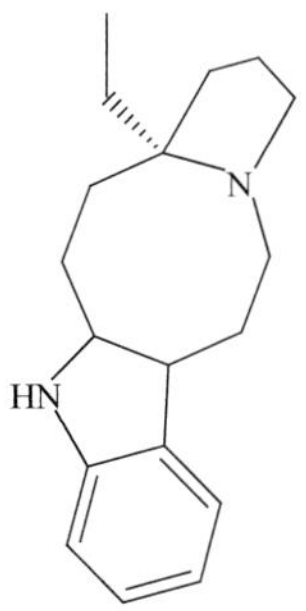

Fig. 185. Structure of quebrachamine.

Actions: Tonic, febrifuge and anti-asthmatic.

Therapeutics: Dyspnoea, emphysema and asthma.

Part used: Bark.

Atropa belladonna L.

Family: Solanaceae.

Common name: Deadly nightshade.

Distribution: Native to Europe, North Africa, and Western Asia.

Botany: A herbaceous perennial herb. It has a fleshy rootstock and grows up to 2 metres. The leaves are tall and ovate. The flowers are bell-shaped, dull purple in colour with green tinges and faintly scented. The fruits are berries.

Chemical composition: Tropane alkaloids including atropine.

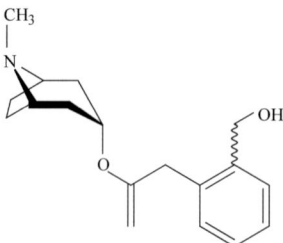

Fig. 186. Structure of atropine.

Actions: Anodyne, muscle relaxant and anti-inflammatory.

Therapeutics: Menstrual problems, peptic ulcer disease, histaminic reaction and motion sickness.

Part used: Leaves.

Atropa acuminata Royle Ex Lindl.

Family: Solanaceae.

Common name: Indian belladonna.

Distribution: Native to the Himalayas.

Botany: Herb up to 1.6 m tall with alternate, ovate-lanceolate acuminate leaves; flowers yellow, 2–2.5 cm long, stamens included.

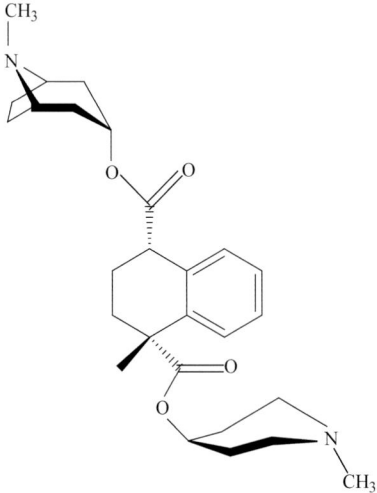

Fig. 187. Structure of belladonnine.

Chemical composition: Tropane alkaloids including atropine and belladonnine.

Actions: Anodyne, muscle relaxant and anti-inflammatory.

Therapeutics: In Ayurvedic medicine, *A. acuminata* is used for the treatment of fever, chicken pox, cold, colitis, conjunctivitis (inflamed eyes) and diarrhoea.

Part used: Leaves.

Brayera anthelmintica Kunth

Syn: *Hagenia abyssinica* (Bruce) J.F. Gmelin.

Family: Rosaceae.

Common name: Cusso, Brayera.

Distribution: Indigenous to north-east Africa.

Botany: A tree native reaching twenty or more feet in height. Leaves are alternate, sessile and interruptedly pinnate, flowers dioecious, small, greenish, turning purple, on hairy peduncles. Leaflets all sessile, the lateral in about 5 pairs with an obliquely cordate base, varying in length to 3 or 4 in; minute leafy lobes are intercalated between the jugæ. Panicles varying in length to 1 ft or more.

Chemical composition: Phloroglucinol derivatives: kosotoxin and protokosin.

Actions: Anthelmintic.

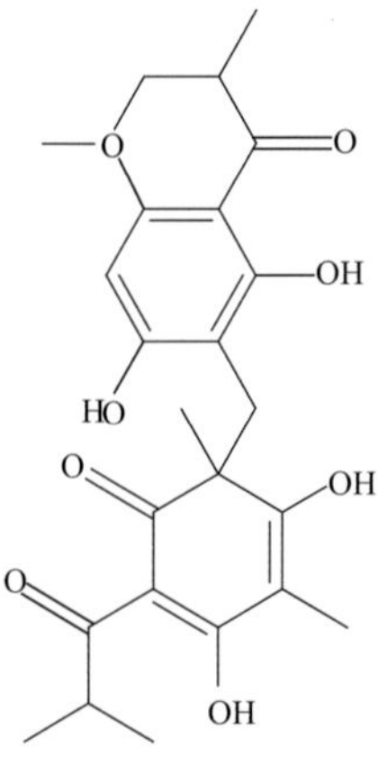

Fig. 188. Structure of kosotoxin.

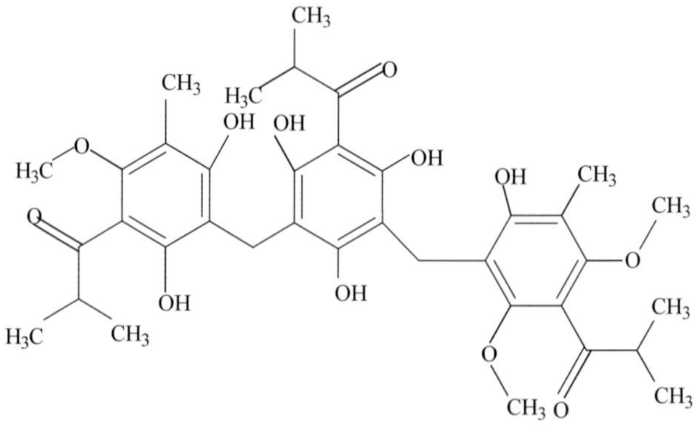

Fig. 189. Structure of protokosin.

Therapeutics: Administered in the form of an infusion for the expulsion of tapeworm. Irritant to mucous membrane; produces nausea, vomiting and colic in large doses. Since its purgative power is often insufficient to expel the head of the parasite, a brisk antibilious cathartic is often necessary (if the bowels do not move within 4 hours).

Part used: Fresh or dried flowers; herb and unripe fruit.

Catha edulis (Vahl) Forssk. ex Endl.

Common name: Khat.

Family: Celastraceae.

Distribution: Native to the Horn of Africa and the Arabian Peninsula.

Botany: *C. edulis* is a shrub or a tree that grows slowly. It attains height of between 1 and 5 m. The leaves are evergreen. The flowers are produced on short axillary cymes. The fruit is capsule having one to three seeds.

Chemical composition: Alkaloids cathine and cathinone.

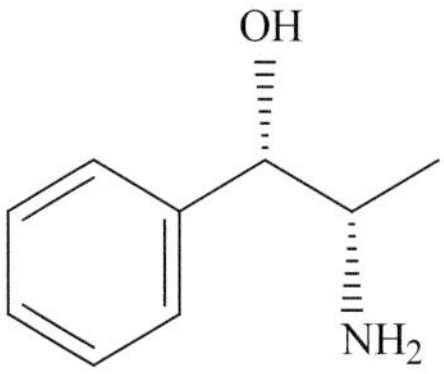

Fig. 190. Structure of cathine.

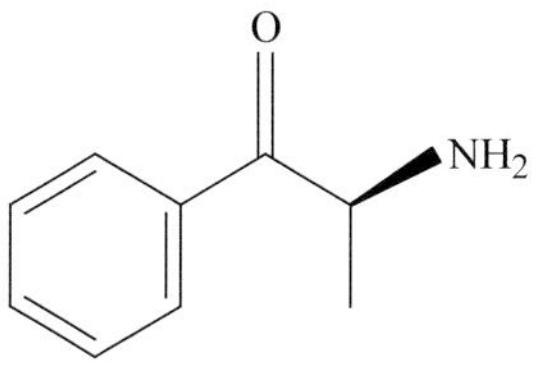

Fig. 191. Structure of cathinone.

Actions: Stimulant.

Part used: Leaves.

Legal status: In 1980, the World Health Organisation classified *C. edulis* as a drug of abuse that can result in mild to moderate psychological dependence. Despite this fact, the World Health Organisation does not consider *C. edulis* to be addictive of a serious nature. The legal status of *C. edulis* may vary from one country to another. Whereas, *C. edulis* falls in the category of a controlled or illegal substance in some countries, in others, it is legal for sale and production. The status of khat in various countries is tabulated below:

	Country	**Status**
1.	Ethiopia	Legal
2.	Somalia	Legal
3.	Kenya	Legal, however, the alkaloids are placed in category C
4.	S. Africa	Protected
5.	China	Illegal
6.	Israel	Legal
7.	Indonesia	Legal
8.	Philippines	Legal
9.	Thailand	Legal
10.	Denmark	Illegal

11.	Finland	Illegal
12.	France	Restricted as a stimulant
13.	Ireland	Controlled
14.	Romania	Illegal
15.	Sweden	Prohibited
16.	Switzerland	Illegal
17.	United Kingdom	Illegal
18.	United States of America	Cathinone is included in Schedule I

Chelidonium majus L.

Common name: Greater celandine.

Family: Papaveraceae.

Distribution: Native to Europe and western Asia and introduced widely in North America.

Botany: *C. majus* is an erect perennial herb reaching 30 to 120 cm height. The leaves are pinnate and have wavy-edged margins. The flowers are yellow in colour. The fruit is a cylindrical capsule. The seeds are small and black.

Chemical composition: Coptisine, allocryptopine, stylopine, protopine, norchelidonine, berberine, chelidonine, sanguinarine, chelerythrine and 8-hydroxydihydrosanguinarine.

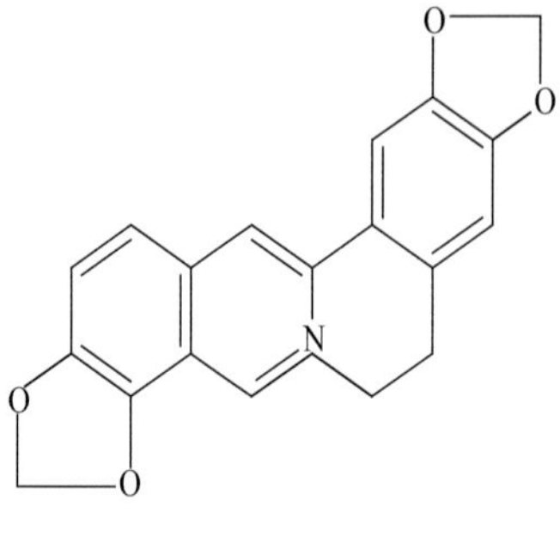

Fig. 192. Structure of coptisine.

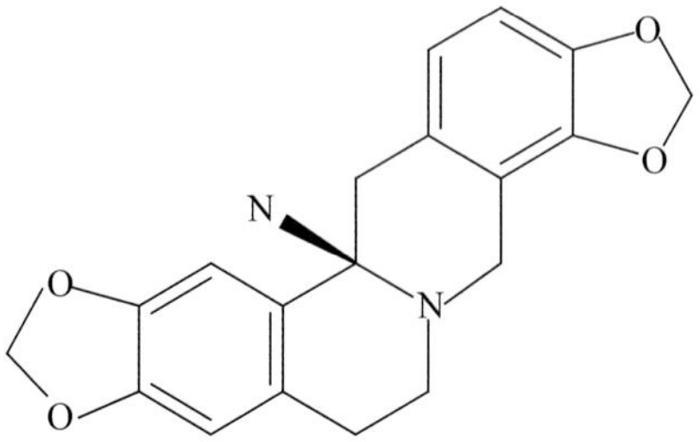

Fig. 193. Structure of stypoline.

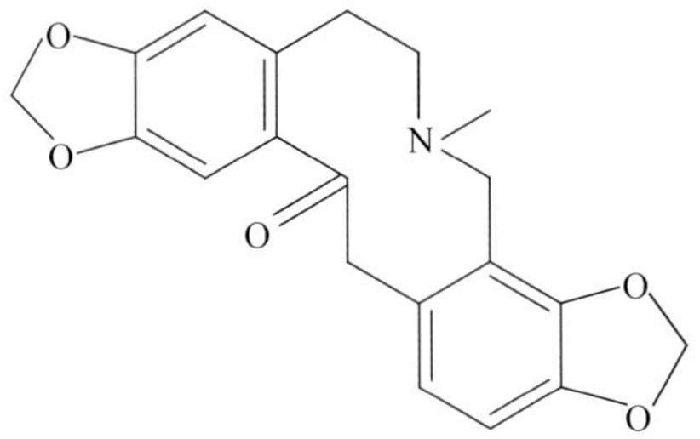

Fig. 194. Structure of protopine.

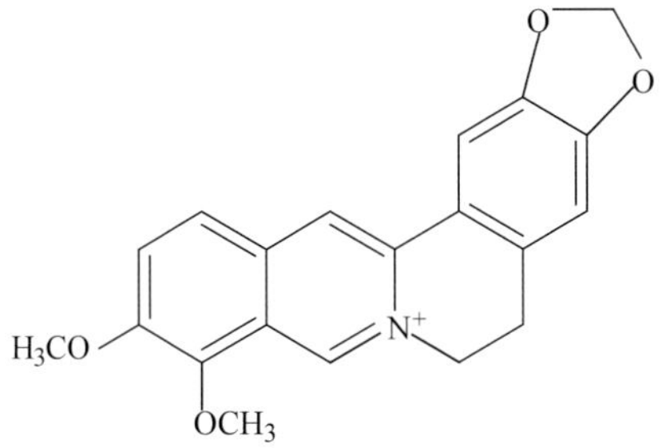

Fig. 195. Structure of berberine.

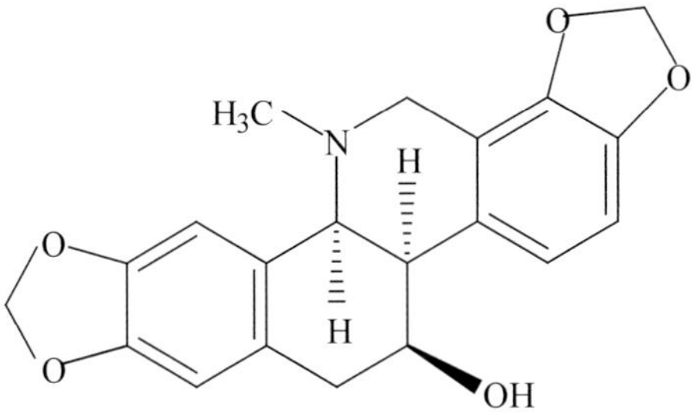

Fig. 196. Structure of chelidonine.

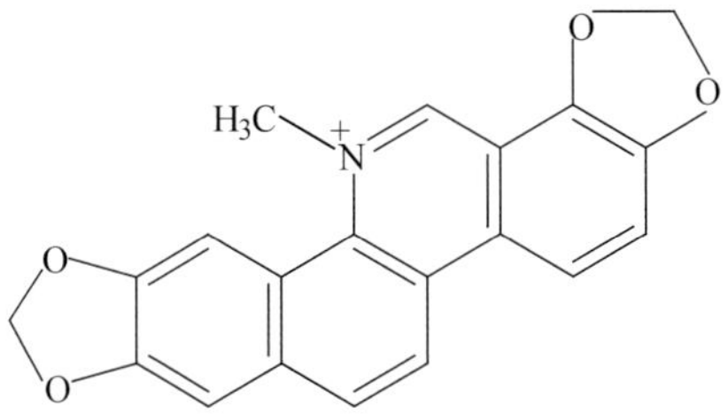

Fig. 197. Structure of sanguinarine.

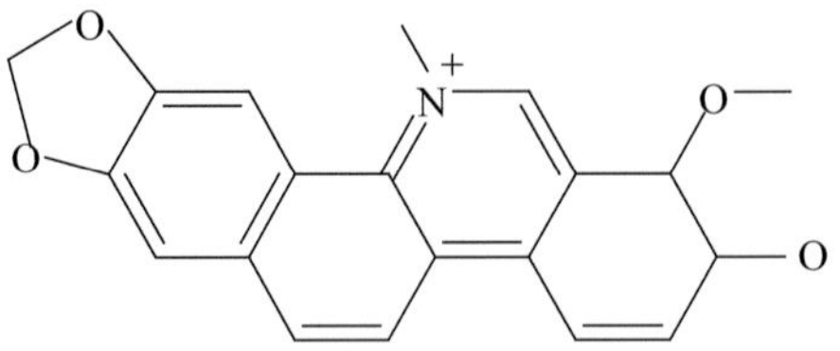

Fig. 198. Structure of chelerythrine.

Actions: Alterative, diuretic and purgative.

Therapeutics: Given in jaundice, eczema and scrofulous diseases.

Parts used: Whole plant.

Chenopodium ambrosioides anthelminticum (L.) A. Gray.

Syn: *Chenopodium anthelminticum* L.

Family: Rubiaceae.

Common name: American wormseed.

Distribution: Waste places from New England to Florida and westward to California.

Botany: *C. anthelminticum* is a much-branched herb growing 2 to 3 feet in height. The leaves are numerous and lance-shaped. The whole plant has a strong, disagreeable odor due to the essential oil.

Chemical composition: Essential oil containing ascaridole, p-cymene, α-terpinene, dihydro-p-cymene and sylvestrene.

Actions: Anthelmintic.

Therapeutics: Ascaridiasis.

Part used: Fruit.

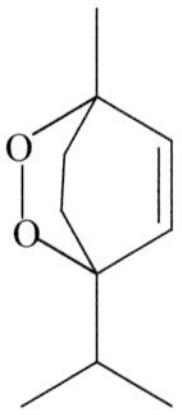

Fig. 199. Structure of ascaridole.

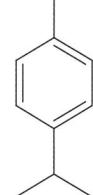

Fig. 200. Structure of p-cymene.

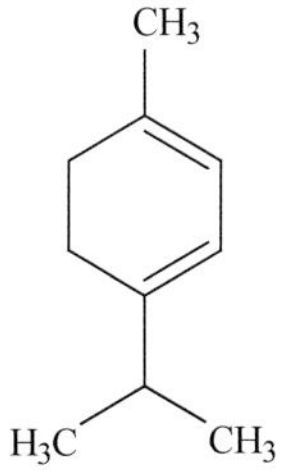

CH₃ structure

Fig. 201. Structure of α-terpinene.

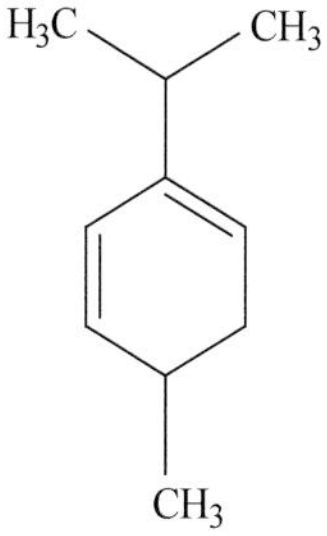

Fig. 202. Structure of dihydro-p-cymene.

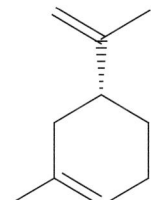

Fig. 203. Structure of sylvestrene.

Cinchona calisaya Weed.

Family: Rubiaceae.

Common name: Peruvian bark.

Distribution: The Central and Southern United States.

Botany: *C. calisaya* is a small tree having large glossy leaves and cymes of fragrant yellow to green or red flowers.

Chemical composition: Quinine, quinidine, cinchonine and chinchonidine.

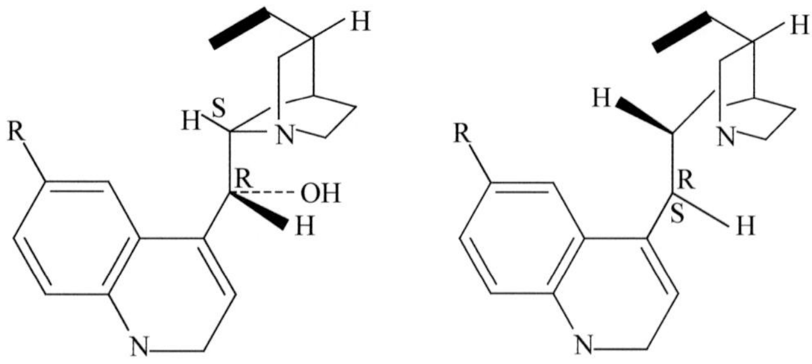

R= OMe, (-) quinine
R= OH. (-) cinchonidine

R= OMe, (+) quinidine
R= OH. (+) cinchonine

Fig. 204. Structure of Cinchona alkaloids.

Actions: Antiperiodic.

Therapeutics: Intermittent fever.

Part used: Bark.

Cinchona ledgeriana

Family: Rubiaceae.

Common name: Peruvian bark.

Distribution: Indigenous to the eastern slopes of the Andes.

Botany: *C. ledgeriana* grow 15–20 metres in height and have large glossy leaves.

Chemical composition: Quinine.

Actions: Antiperiodic.

Therapeutics: Intermittent fevers.

Part used: Bark.

Cinchona succirubra Pav

Family: Rubiaceae.

Common name: The red quinine tree.

Distribution: South America, but cultivated in India, Java and Ceylon.

Botany: *C. succirubra* is an evergreen tree.

Chemical composition: Quinine in high amounts.

Actions: Antiperiodic.

Therapeutics: Malaria and influenza.

Part used: Bark.

Cinchona micrantha Ruiz and Pav.

Family: Rubiaceae.

Distribution: South America.

Botany: *C. micrantha* is an evergreen tree.

Chemical composition: Cinchonine.

Actions: Antiperiodic.

Therapeutics: Malaria and influenza.

Part used: Bark.

Claviceps purpurea (Fr.) Tul

Common name: Ergot.

Family: Clavicipitaceae.

Distribution: Northern temperate regions.

Botany: *C. purpurea* grows on rye and related plants. An ergot kernel, called a sclerotium, develops when a spore of fungal species of the genus *Claviceps* infects a floret of flowering grass or cereal.

Chemical composition: 0.15–0.5% alkaloids. Three classes of ergot alkaloids: the ergotamine group (ergotamine, ergosine and the corresponding isomers) the ergotoxine group (ergocornine, ergocristine, ergokryptine and their isomers) and ergotamine group (ergometrine and its isomer).

Actions: Ecbolic and styptic.

Therapeutics: It finds application in anti-migraine prescriptions. Ergometrine is used as an oxytocic, and is injected during the final stages of labour and immediately following childbirth, especially if hemorrhage occurs. Bleeding is reduced because of its vasoconstrictor effects, and it is valuable after Caesarean operations.

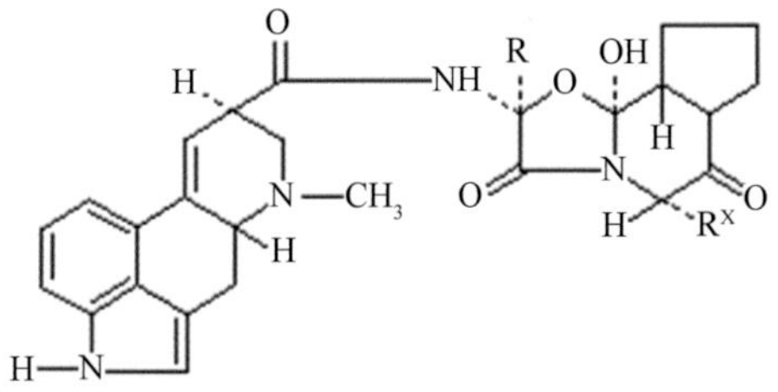

Fig. 205. Chemical structure of ergotamine.

Part used: Sclerotium.

Toxicity: Alkaloids can cause ergotism in humans and other mammals who consume grains contaminated with its sclerotium.

Clematis armandii Franch.

Common name: Evergreen clematis.

Distribution: Native to China and North Burma.

Botany: An evergreen herb with fragrant flowers.

Chemical composition: Lignans: (7R,8S)-9-acetyl-dehydrodiconiferyl alcohol, (7R,8S)-dehydrodiconiferyl alcohol, erythro-guaiacylglycerol-β-coniferyl ether, and threo-guaiacylglycerol-β-coniferyl ether.

Therapeutics: Embolism and rheumatism.

Part used: Roots.

Clematis montana Buch.-Ham. ex DC.

Common name: Himalayan clematis.

Distribution: Native to the mountain areas of Asia from Afghanistan to Taiwan.

Botany: A woody climber.

Chemical composition: Triterpenic bisglycosides: clemontanoside B–F, and lignans.

Therapeutics: The leaves are used in the treatment of skin diseases and seeds are purgative.

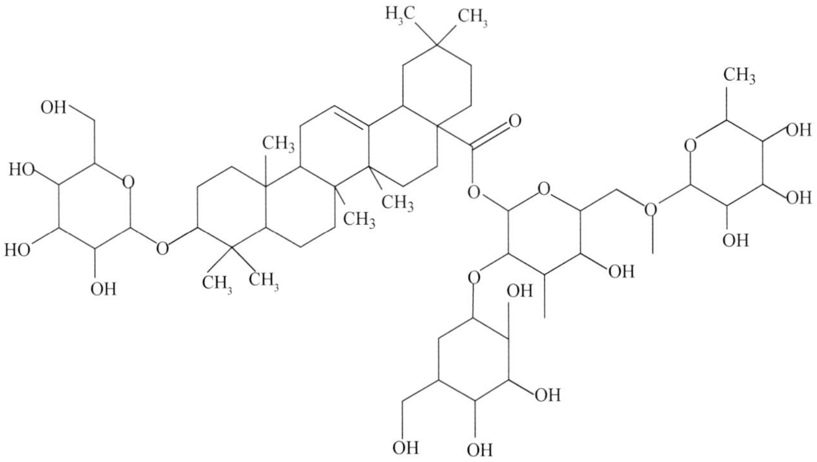

Fig. 206. Structure of clemontanoside B.

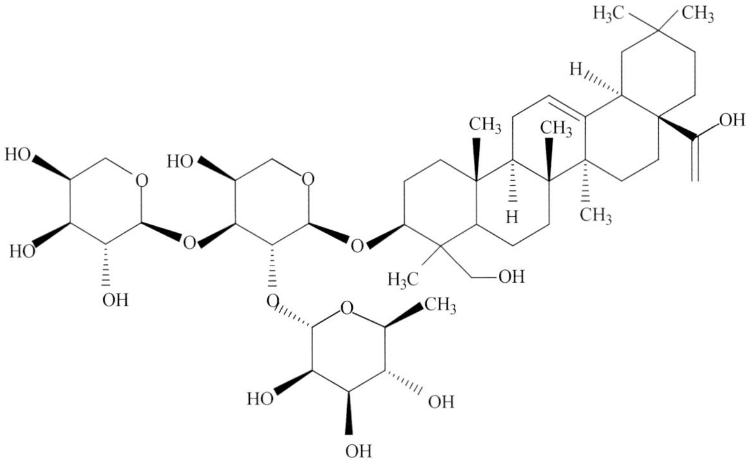

Fig. 207. Structure of clemontanoside C.

Cocculus indicus L.

Syn: *Anamirta cocculus* (L.) Wight and Arn.

Family: Menispermaceae.

Common name: Fishberry.

Distribution: Native to China and Japan.

Botany: *C. indicus* is a large-stemmed plant. The colour of the bark is gray with white wood. The flowers are aromatic, small and yellowish-white in colour. The fruit is a drupe.

Chemical composition: Alkaloids: berberine, palmatine, magnoflorine and columbamine (roots) and menisperimine and paramenispermine (seed shells) and neutral principle: picrotoxin (in seeds).

Actions: Parasiticidal.

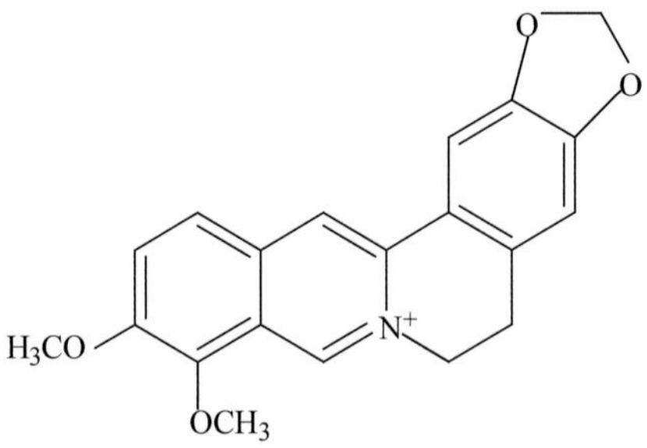

Fig. 208. Structure of berberine.

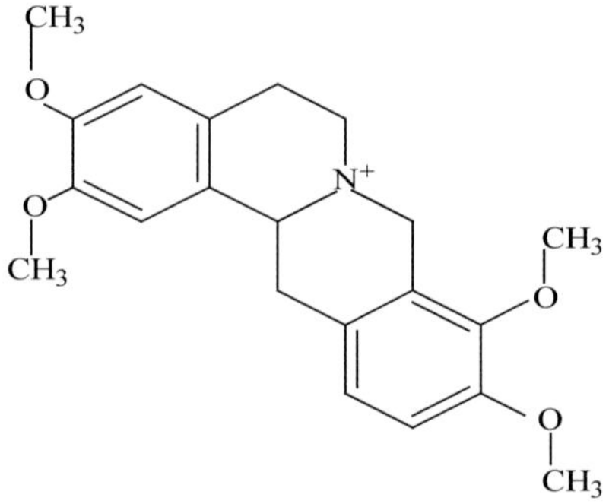

Fig. 209. Structure of palmatine.

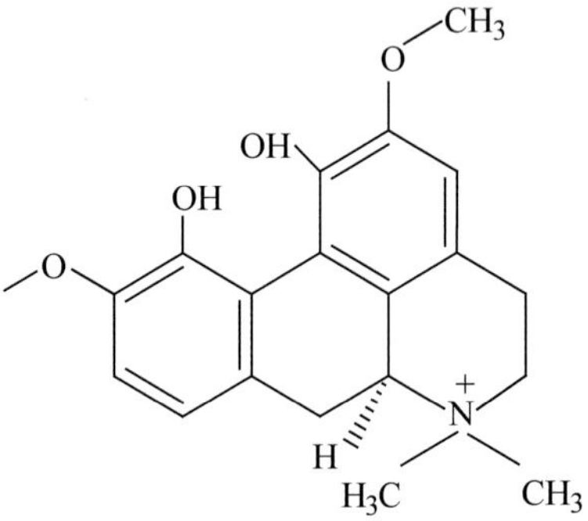

Fig. 210. Structure of magnoflorine.

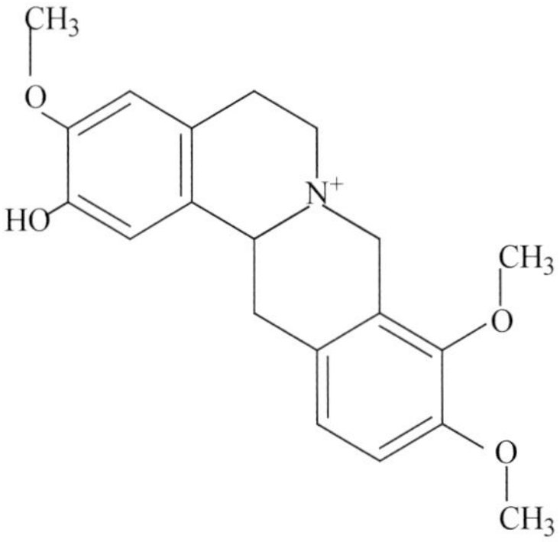

Fig. 211. Structure of columbamine.

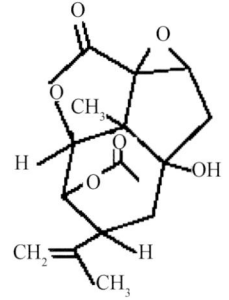

Fig. 212. Structure of picrotoxin.

Therapeutics: The powdered barriers are occasionally used in the treatment of pediculosis. Picrotoxin has been used but seldom in the treatment of hectic fever as it is a powerful convulsant.

Parts used: Berries.

Cocculus laurifolius DC.

Family: Menispermaceae.

Common name: Laurel-leaved snail tree.

Distribution: Native to China and Japan.

Botany: A medium sized tree grow in to the height of 40 to 60 feet. Leaves are simple, large and green in colour. Flowers are white with yellow spots in the spring season. Fruit is an elongated capsule.

Chemical composition: Alkaloids: cocculine (sinomenine), cocculitine, isoboldine, norisoboldine, coclafine and coccuvine.

Actions: Diuretic and vermifuge.

Part used: Roots.

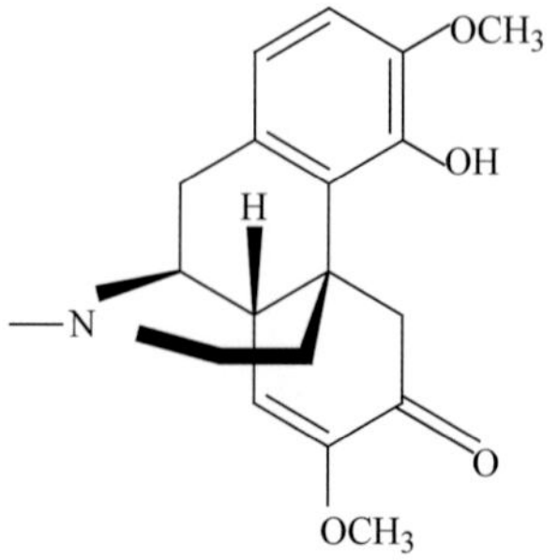

Fig. 213. Structure of cocculine.

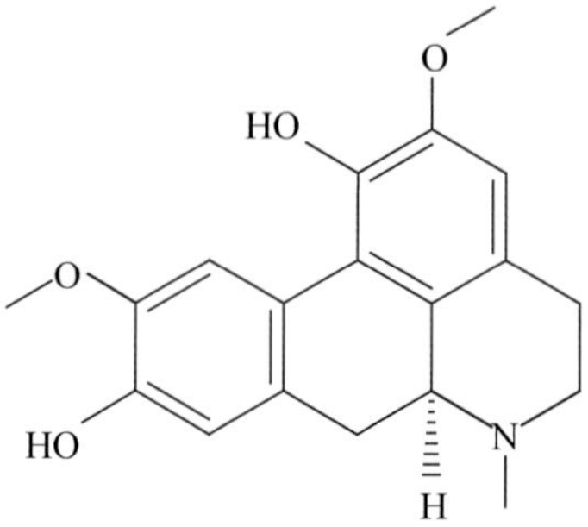

Fig. 214. Structure of isoboldine.

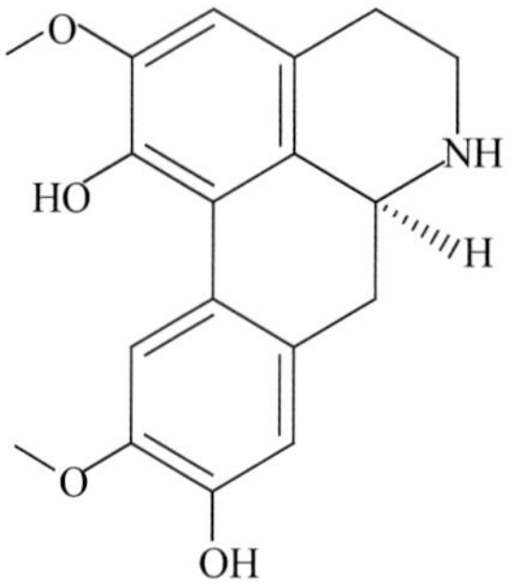

Fig. 215. Structure of norisoboldine.

Cocculus orbiculatus (L.) DC.

Family: Menispermaceae.

Common name: Queen Coralbead.

Distribution: Eastern India to Java.

Botany: A slender woody climbing vine. The shape of the leaves can vary from linear-lanceolate to broad-elliptic to broad-ovate. The flowers are golden yellow in colour. Fruit is pea sized and blue or black on maturing.

Chemical composition: Protoberberine alkaloids.

Actions: anodyne, antiphlogistic, antirheumatic, carminative, depurative, diuretic and vermifuge.

Therapeutics: Rheumatic arthritis, oedema and oliguria.

Part used: Roots.

Colchicum autumnale L.

Family: Colchicaceae.

Common name: Meadow saffron.

Distribution: Native to the Great Britain and Ireland.

Botany: A small perennial herb growing upto 10 to 40 cm. It flowers typically in the autumn after the disappearance of the leaves. Leaves are lanceolate, dark green in colour and, shiny. Flowers are showy, pink, purple to white in colour produced from an underground bulb.

Chemical composition: Colchicine is present in the flowers (0.1 to 0.8% in fresh flowers; up to 1.8% in dried flowers), in the seeds (0.2 to 0.8%) in the bulb (0.4 to 0.6%). The leaves contain very low amounts of colchicine. The other toxins present, which are closely related to colchicine, includes demecolcine (desacetylmethylcolchicine), desacetylthiocolchicine, colchicoside, and demethyl desacetylcolchicine.

Actions: Alterative, aphrodisiac, carminative and laxative.

Therapeutics: Gout, rheumatism and disorders of liver and spleen.

Part used: Corms.

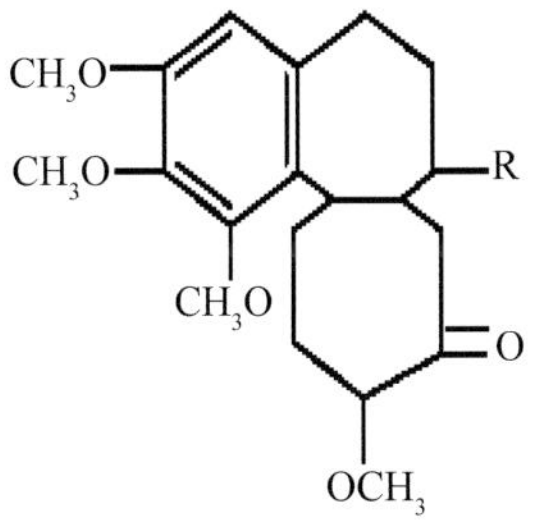

Fig. 216. Structure of colchicine.

Conium maculatum L.

Common name: Poison hemlock.

Family: Apiaceae.

Common name: Native to Europe and North Africa.

Distribution: Eastern India to Java.

Botany: A biennial herb growing up to 1.5–2.5 m tall. The stem is smooth and green in colour. All parts of the plant are hairless. The leaves are two- to four-pinnate. The flowers are small, white in colour and clustered in umbels.

Chemical composition: Alkaloids: conine, gamma-coniceine, N-methylconiine, conhydrine and pseudoconhydrine.

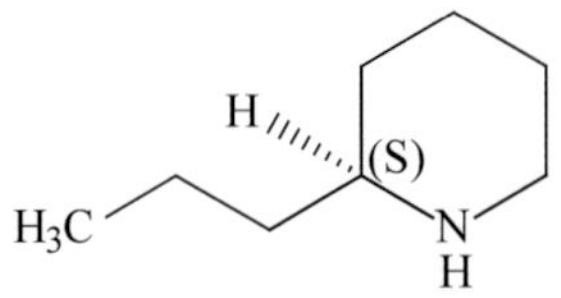

Fig. 217. Structure of conine.

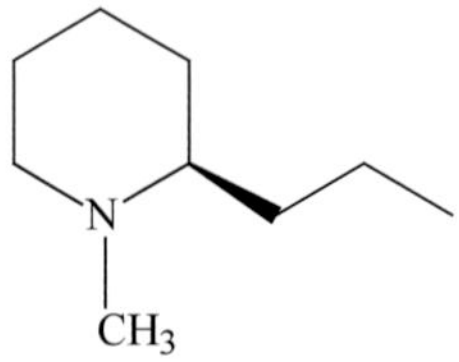

Fig. 218. Structure of N-methylconiine.

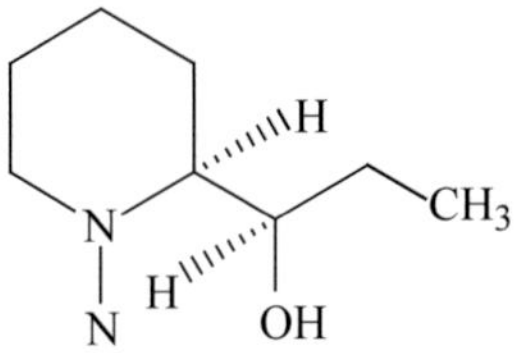

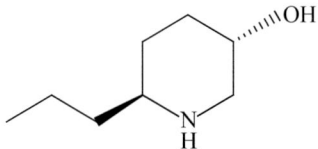

Fig. 219. Structure of conhydrine.

Fig. 220. Structure of pseudoconhydrine.

Actions: Neuromuscular blocking agent.

Part used: Flower.

Convallaria majalis L.

Common name: Lily of the valley.

Family: Fabaceae.

Distribution: Native of Europe.

Botany: *C. majalis* is a perennial herb. It forms colonies from rhizomes. The stems contains one or two leaves 10–25 cm long. The flowering stem has two leaves and a raceme of 5–15 flowers on the stem apex. The fruit is a small and orange-red coloured berry having brown or white seeds.

Chemical composition: Cardiac glycosides (convallarin, convallamarin, convallatoxin, convallatoxol, convallotoxoloside, neoconvalloside, locundioside and many more).

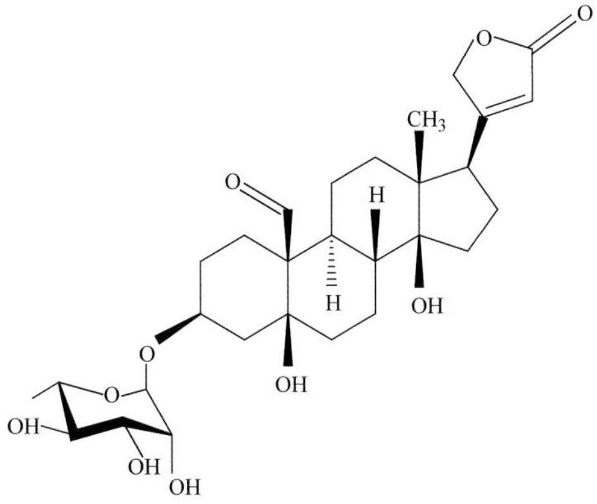

Fig. 221. Structure of convallatoxin.

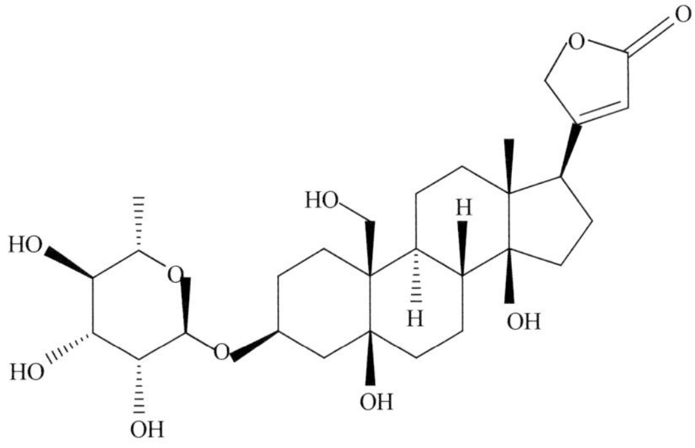

Fig. 222. Structure of convallatoxol.

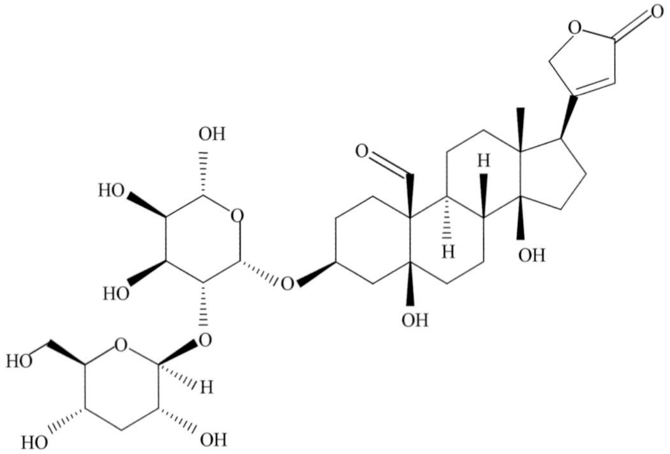

Fig. 223. Structure of neoconvalloside.

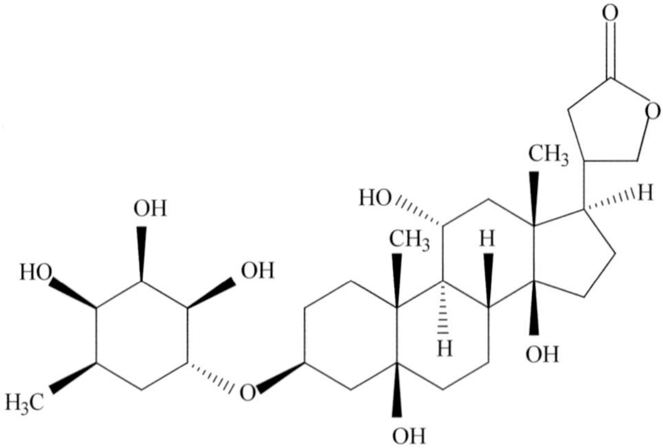

Fig. 224. Structure of locundioside.

Actions: Cardiac tonic.

Therapeutics: Cardiac anasarca.

Part used: Bulbs.

Crotalaria fulva Stuart and Bras

Syn: *Crotalariaberteroana* DC.

Family: Fabaceae.

Common name: Whiteback.

Distribution: Australia.

Botany: A herbaceous plant.

Chemical composition: Pyrrolizidine alkaloid: fulvine (Schoental 1963).

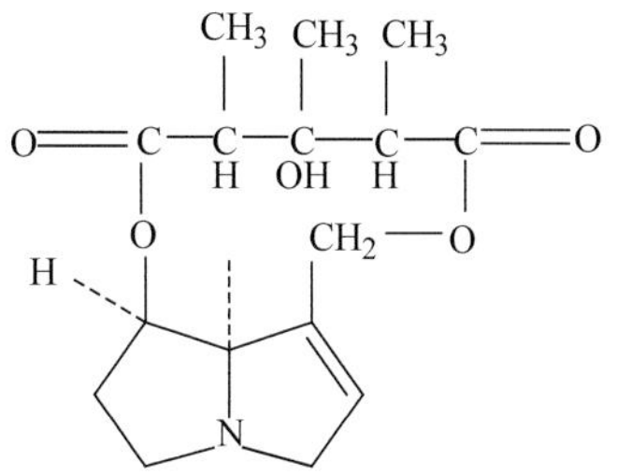

Fig. 225. Chemical structure of fulvine.

Therapeutics: *C. fulva* is used in the West Indies for the preparation of bush tea, which is used by the indigenous community for medicinal value.

Toxicity: Veno-occlusive Lesions have been reported in livers of rats fed with *C. fulva* (McLean et al. 1964). The oral or systemic administration of fulvine to rats leads to right ventricular hypertrophy accompanied by thickening of the pulmonary trunk and medial hypertrophy of muscular pulmonary arteries (Kay et al. 1971; Wagenvoort et al. 1974).

Curcurbita maxima Duch

Family: Cucurbitaceae.

Common name: Pumpkin.

Distribution: Native to South America.

Botany: *C. maxima* is annual herbaceous climber. The stems, more or less prickly. Leaves are simple, alternate, and shallowly to deeply lobed. Fruits are relatively large.

Chemical composition: Essential oil, sterols and amino acid, curcurbitin.

Actions: Anti-inflammatory, antipyretic, diuretic, tonic, and vermifuge.

Therapeutics: Boils, ulcer, stomach pain, haemorrhage and venomous insect bites.

Parts used: Fruit and seeds.

Datura innoxia Mill.

Common name: Recurved thorn-apple.

Family: Solanaceae.

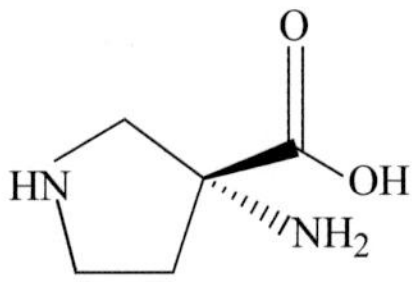

Fig. 226. Chemical structure of curcurbitin.

Distribution: Native to Central and South America, and introduced in Africa, Asia, Australia and Europe.

Botany: *D. innoxia* is an annual shrub reaching a height of 0.6 to 1.5 metres. The stems and leaves have a covering of short and soft grayish hairs. The leaves are elliptic and entire-edged with pinnate venation.

Chemical composition: Tropane alkaloids.

Actions: Anticholinergic, bronchodilator, mydriatic and narcotic.

Therapeutics: Bronchial asthma.

Parts used: Leaves and seeds.

Toxicity: All parts of *D. innoxia* contain dangerous levels of poison and may be fatal if ingested by humans and other animals, including livestock and pets.

Datura stramonium L.

Common name: Jimson weed.

Family: Solanaceae.

Distribution: Native to Mexico and widely naturalised.

Botany: *D. stramonium* is an erect, annual, growing up to 60 to 150 cm. The root is long, thick, fibrous and white. The stem is erect, smooth and pale yellow-green in colour. The stem forks off repeatedly into branches. Each fork forms a leaf and a single, erect flower. The leaves are, smooth, toothed, and undulated.

Chemical composition: Alkaloids (daturine (hyoscyamine), norhyosciamine and meteloidine), chlorogenic acid, fixed oil and volatile oil.

Actions: Anticholinergic, bronchodilator, mydriatic and narcotic.

Therapeutics: Bronchial asthma.

Parts used: Leaves and seeds.

Toxicity: All parts of *D. stramonium* contain dangerous levels of poison and may be fatal if ingested by humans and other animals, including livestock and pets.

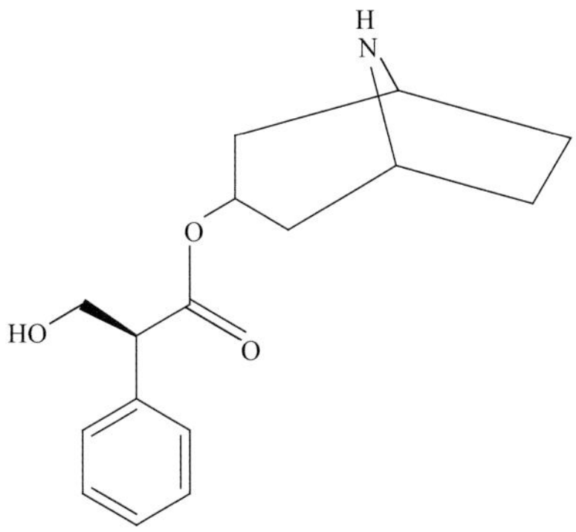

Fig. 227. Chemical structure of norhyosciamine.

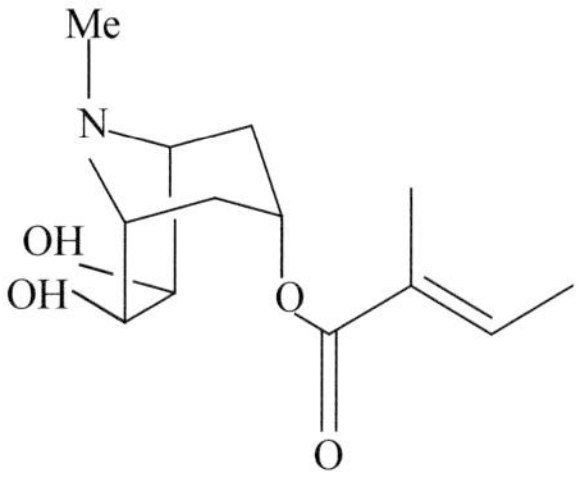

Fig. 228. Chemical structure of meteloidine.

Digitalis lanata Ehrb.

Common name: Grecian foxglove.

Family: Scrophulariaceae.

Distribution: Eastern Europe.

Botany: *D. lanata* grows up to 0.6 meters. The elongated leaves are mid–green, woolly, veined, and covered with white hairs on the underside. The flowers are tubular, bell shaped and creamy-white in color.

Chemical composition: The active principles are glycosides (digoxin, gitoxin and digitoxin).

Actions: Cardiac tonic.

Therapeutics: Congestive cardiac failure.

Parts used: Leaves.

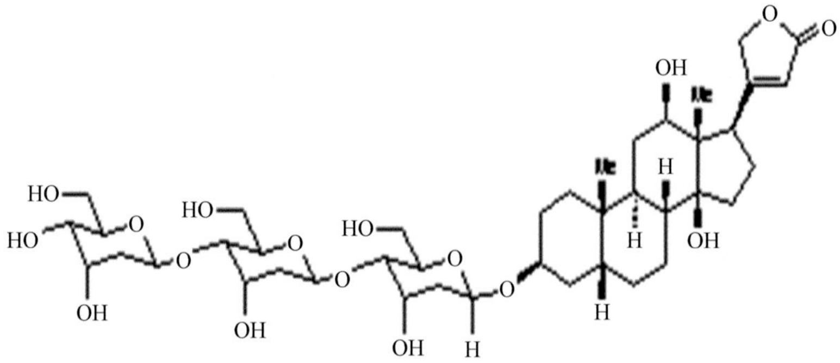

Fig. 229. Chemical structure of digoxin.

Digitalis purpurea L.

Common name: The purple foxglove.

Family: Scrophulariaceae.

Distribution: Native to temperate Europe.

Botany: A biennial or perennial herb. The leaves are spirally arranged, simple, covered with gray-white pubescent and glandular hairs. The flowers are arranged in a showy, terminal, elongated cluster. The flowers are typically purple in colour. The fruit is a capsule with 0.1–0.2 mm seeds.

Chemical composition: The active principles are glycosides (digitoxin, gitoxin and gitalin). Digitoxin is sparingly soluble in water but soluble in alcohol. It is the main active constituent of infusions. Gitoxin in pure state is insoluble in water. Saponins are present in leaves and seeds but exert no digitalis effect. Saponins affect the solubility of the cardiac glucosides. Glycosides are broken down by hydrolysis into aglycone and sugar molecule. The activity of the glucoside depends on aglycone portion which has a steroid nucleus. The sugar portion increases solubility and cell permeability of glycone.

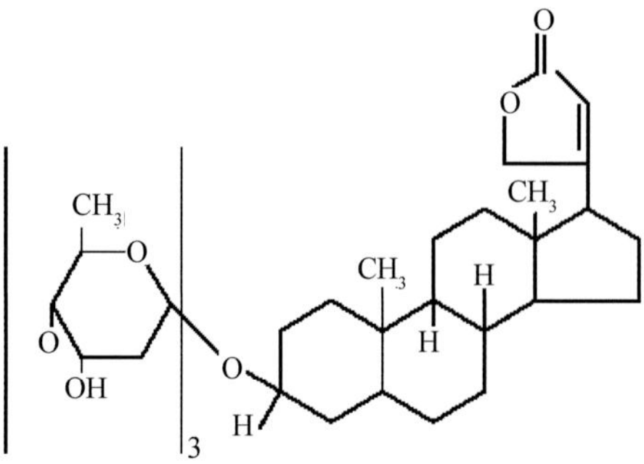

Fig. 230. Chemical structure of Digitoxin.

Stoll and Kries showed that *D. purpurea* glycosides are divided into two groups:

(A) $C_{47}H_{74}O_{18}+H_2O=C_{47}H_{64}O_{13}+C_6H_{12}O_6$
    Glycoside-A   Digitoxin Glucose

(B) $C_{47}H_{74}O_{19}+H2O=C_{47}H_{64}O_{14}+C_6H_{12}O_6$
    Glycoside-B   Gitoxin Glucose

Aglycone for Glycoside-A is digitoxigenin and for Glycoside-B is gitoxigenin and gitaligenin.

Actions: Cardiac tonic.

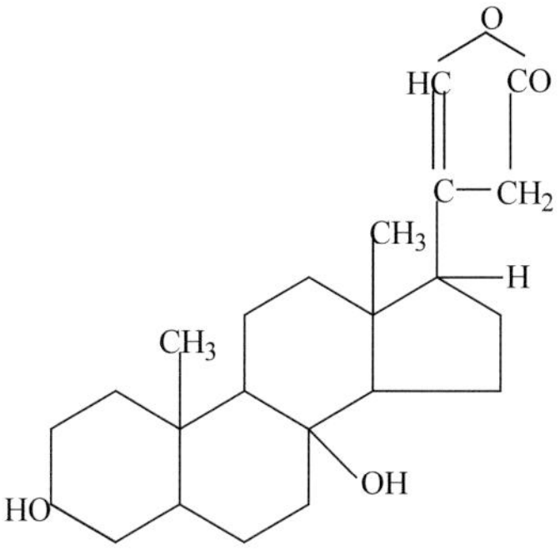

Fig. 231. Structure of digitoxigenin.

Therapeutics: Congestive cardiac failure.

Parts used: Leaves.

Toxicology of digitalis: *D. purpurea* is a cumulative poison. The main symptoms of poisoning include vomiting, diarrhoea, bradycardia, anuria, headache, nausea and coupling of beats.

Treatment: To remove the poison, stomach is washed and analeptics are given. Bed rest is of prime importance.

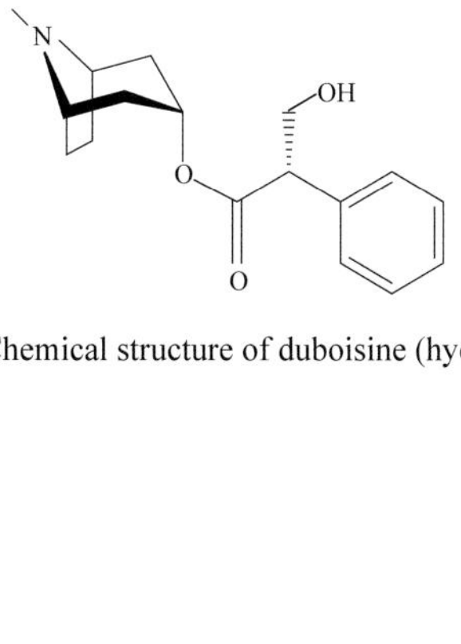

Fig. 232. Chemical structure of duboisine (hyoscyamine).

Duboisia leichardtii (F. Muell.) F. Muell.

Family: Solanaceae.

Distribution: Cultivated in Australia and Japan.

Botany: A tall shrub or a small tree growing upto 12 metres tall.

Chemical composition: Tropane alkaloids including duboisine (hyoscyamine).

Actions: Hypno-sedative.

Therapeutics: In Homeopathy, the plant is used in eye-diseases.

Duboisia myoporoides R.Br.

Common name: Corkwood tree.

Family: Solanaceae.

Distribution: Eastern Australia.

Botany: A shrub or tree having thick and corky bark. The leaves are obovate to elliptic in shape. The small white flowers are produced in clusters. The berries are globose and purple-black in colour.

Chemical composition: Alkaloids including hyoscine (scopolamine).

Actions: Hypno-sedative and mydriatic.

Therapeutics: In Homeopathy, the plant is used in treating diseases related to vision and paralysis.

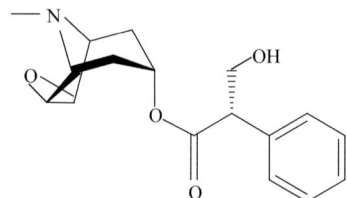

Fig. 233. Chemical structure of hyoscine.

Dryopteris filix-mas (L.) Schott

Common name: Worm Fern, European male fern.

Family Dryopteridaceae.

Distribution: Grown as an ornamental fern in gardens.

Botany: A semi-evergreen fern. The bipinnate leaves consist of 20–35 pinnae on each side of the rachis. The stalks are covered with orange-brown scales.

Chemical constituents: Phloroglucinols corresponds to 2% and is also termed as filicin. Major phloroglucinols includes filicic acid, which is probably the most active. Filicic acid contains not less than 1.5% of filicin. Liquid extract of male fern contains 25% w/w of filicin.

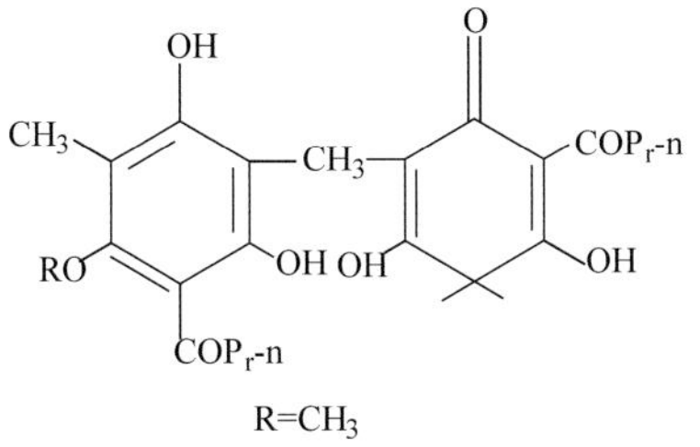

R=CH₃

Fig. 234. Chemical structure of filicic acid.

Actions: Anthelmintic.

Therapeutics: Preliminary preparation of *D. filix-mas* with milk diet and alba mixture for two days is followed by the drug in 3 divided doses in succession on empty stomach in the morning. This is yet again followed by a magnesium sulphate purgative. The head of the tapeworm is looked for in the stools passed. Castor oil must not be used because oils aid the absorption of filicic acid which is toxic.

Contraindications: Alcoholism, pregnancy and in advanced diseases of the heart, liver and kidneys.

Toxicity: Hepatotoxic and nephrotoxic.

Substitutes: *Asplenium Filix-foemina* (Bernh.), *Aspidium oreopteris* (Sw.), and *A. spinulosum* (Sw.) and *D. juxtaposita* Christare used as substitutes for European male fern.

Ecballium elaterium (L.) A. Rich.

Common name: Squirting cucumber.

Family: Solanaceae.

Distribution: Native to Europe, northern Africa, and temperate areas of Asia.

Botany: *E. elaterium* is a hairy vine. The leaves are palmately lobed. The fruit is ovoid, fleshy, approximately 4 cm in length. The unripe fruit is of a pale green color.

Chemical composition: Cucurbitacin B and sterols.

Actions: Anti-inflammatory and analgesic.

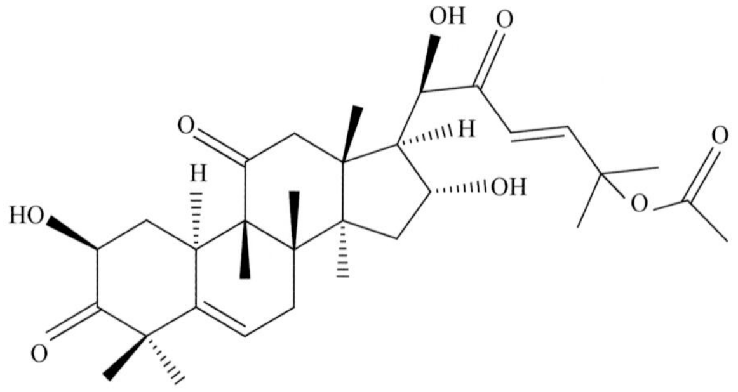

Fig. 235. Chemical structure of cucurbitacin B.

Therapeutics: Fever, cancer, liver disorders, jaundice, constipation, hypertension, dropsy and rheumatic diseases.

Parts used: Fruits.

Toxicity: All parts of *E. elaterium* are toxic, particularly the ovoid green fruits. Progression of the inflammation can cause conjunctival irritation, corneal edema and erosions, sore throat, dysphagia, drooling, dyspnea, or respiratory distress secondary to upper airway edema. Obstruction of the upper airway is a potentially fatal complication of the nasal installation of undiluted juice from *E. elaterium.*

Embelia ribes Burm. f.

Common name: False black pepper.

Family: Myrsinaceae.

Distribution: Throughout India.

Botany: A climbing shrub. The root is brownish-grey in colour and hairy. The stem is whitish grey in colour with protruded lentils. The leaves are elliptic, lanceolate and entirely glabrous. Flowers are white-yellow in color and small. The fruit is berry.

Chemical composition: Embelin, vilangine, homoembelin, christembine (alkaloid) and quercitol.

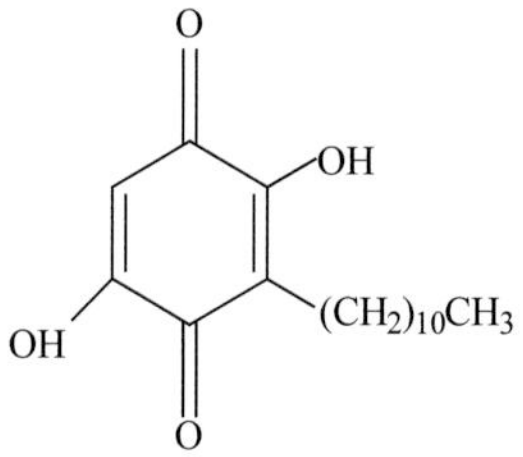

Fig. 236. Chemical structure of embelin.

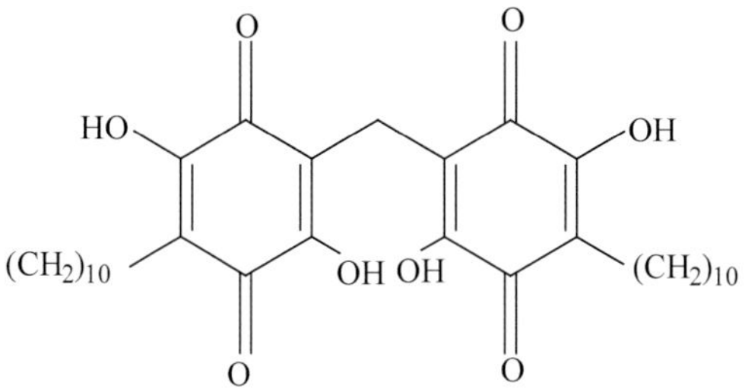

Fig. 237. Chemical structure of vilangine.

Actions: Acrid, astringent, anthelmintic, antifertility, antioestrogenic, carminative, digestive, laxative, soothing, stimulant, stomachic, and thermogenic.

Therapeutics: Abdominal disorders, skin fungal infections, flatulence, constipation, indigestion, headache, haemorrhoids, lung diseases, obesity, piles, pneumonia, mouth ulcers, toothache and sore throat.

Parts used: Fruits, dried bark and roots.

Embelia robusta Roxb.

Syn: *Ribesoides robustum* (Roxb.) Kuntze.

Family: Myrsinaceae.

Distribution: Outer Himalayas in India and Pakistan.

Botany: A large scandent shrub. Leaves are oblong-ovate to elliptic-oblong and very shortly petioled. Racemes are short as compared to the leaves. Flowers greenish-yellow to whitish in colour. Fruit is hardly pulpy.

Chemical composition: Embelin.

Actions: Anthelmintic and cathartic.

Therapeutics: Dried bark and roots are used in treating dental pain.

Parts used: Fruits, dried bark and roots.

Ephedra distachya L.

Family: Ephedraceae.

Common name: Somlata.

Distribution: Southern and central Europe and parts of western and central Asia.

Botany: An evergreen shrub growing about 25–50 cm high. Flowers blossom from April to July and fruits September to October.

Chemical composition: Alkaloids, ephedrine.

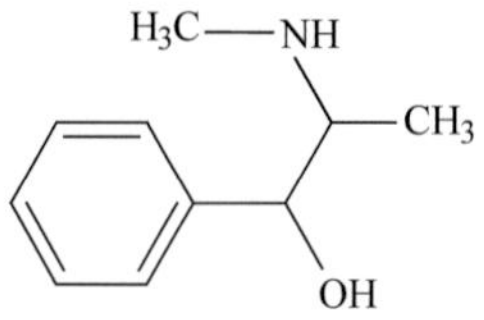

H₃C—NH

—CH₃

OH

Fig. 238. Structure of ephedrine.

Actions: Stimulant and cardiac tonic.

Therapeutics: Asthma.

Parts used: Dried branches.

Ephedra equisetina Bunge

Family: Ephedraceae.

Common name: Bluestem joint fir.

Distribution: Eastern Asia—Northern China.

Botany: An evergreen shrub growing about 1.5 m by 1 m. Flowers bloom by January. The flowers are dioecious.

Chemical composition: Alkaloids, ephedrine and ephedroxane.

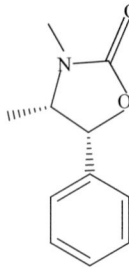

Fig. 239. Structure of ephedroxane.

Actions: Bronchodilator.

Therapeutics: Asthma.

Parts used: Dried branches.

Ephedra intermedia Schrenk et Mey.

Family: Ephedraceae.

Common name: Bluestem joint fir.

Distribution: Temperate Pangi Hills, Kanawa and Kashmir.

Botany: *E. intermedia* grows to 1 metre. The strobili are dioecious, either male or female on any one plant, so both male and female plants are needed for seeds.

Chemical composition: Alkaloid, ephedrine.

Actions: Bronchodilator.

Therapeutics: Aqueous extract is used for treating bronchial asthma.

Parts used: Dried branches.

Ephedra gerardiana Wall. Ex Stapf

Family: Ephedraceae.

Common name: Gerard's Jointfir.

Distribution: Temperate and Alpine Himalayas.

Botany: *E. gerardiana* is a perennial small shrub composed primarily of fibrous stalks, generally about 8 inches though sometimes growing to 24 inches in height, with small, yellow flowers followed by round, red, edible fruits.

Chemical composition: Alkaloids (ephedrine {0.3%} and pesudoephedrine).

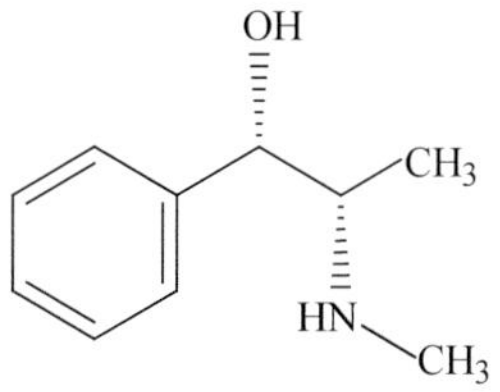

Fig. 240. Structure of pesudoephedrine.

Actions: Bronchodilator.

Therapeutics: It is used in the treatment of hysteria, nocturnal enuresis, bronchial asthma, narcolepsy, dysmenorrhoea and commoncold.

Parts used: Dried branches.

Ephedra sinica Stapf

Family: Ephedraceae.

Common name: Chinese ephedra or Ma Huang.

Distribution: Northeastern China and Mongolia.

Botany: An evergreen shrub.

Chemical composition: Alkaloid, ephedrine and calcium oxalate.

Actions: Bronchodilator.

Therapeutics: It is used in the treatment of hysteria, nocturnal enuresis, bronchial asthma, narcolepsy, dysmenorrhoea and common-cold.

Parts used: Dried branches.

Erysimum canescens Roth

Family: Brassicaceae.

Distribution: Australia.

Botany: A herbaceous plant.

Chemical composition: Cardiac glycosides: helveticoside, erycanoside and erysimine.

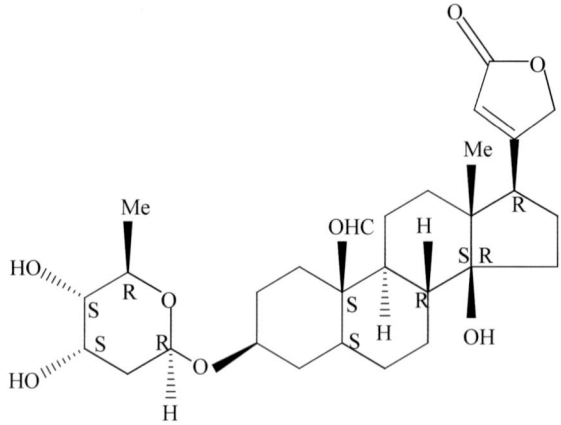

Fig. 241. Chemical structure of helveticoside.

Gelsemium sempervirens (L.) J.St.-Hil.

Common name: Yellow jessamine.

Family: Gelsemiaceae.

Distribution: Native to subtropical and tropical America.

Botany: *G. sempervirens* grow to 3–6 m high. The leaves are evergreen, lanceolate, 5–10 and dark green in colour. The flowers are borne in clusters. The individual flowers are yellow in colour. The flowers have a strong scent.

Chemical composition: Alkaloids: gelsemine and gelseminine.

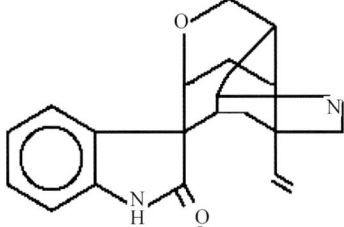

Fig. 242. Chemical structure of gelsemine.

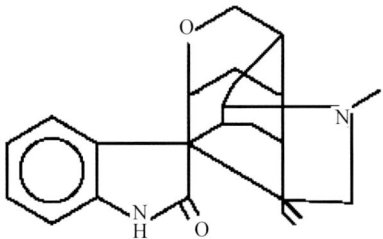

Fig. 243. Chemical structure of gelseminine.

Actions: Convulsant.

Therapeutics: Measles, neuralgic otalgia, tonsillitis, esophagitis, dysmenorrhea, muscular rheumatism, and headaches.

Part used: Rhizomes.

Holarrhena antidysenterica Wall.

Family: Apocynaceae.

Common name: Kurchi.

Distribution: Northeastern China and Mongolia.

Botany: A deciduous shrub or a small tree growing up to 3 m. The stem has several branches. Leaves are arranged in opposite fashion and are acuminate. White flowers are arranged on corymb like cymes. The flowers have oblong petals rounded at the tip.

Chemical composition: Alkaloids (conessine, holarrhenine, kurchine, kurchicine conkurchine, conesimine, holarrhine, holarrhimine, conimine, norconessine, isoconessimine and antidysentericine) gum resin (lettoresinol-A and lettoresinol-B) and foul smelling oil (in seeds).

Actions: Antiprotozoal.

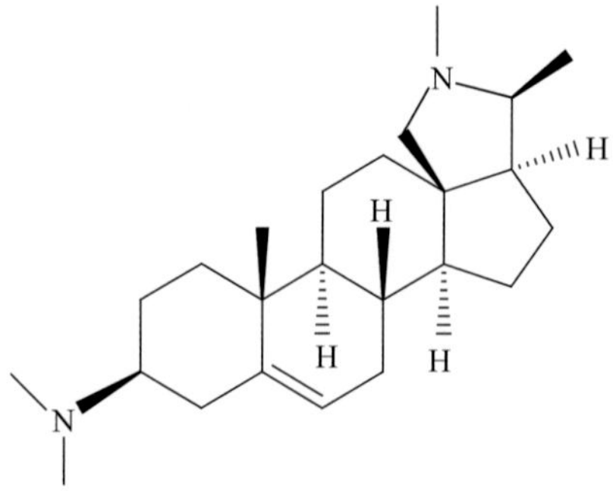

Fig. 244. Structure of conessine.

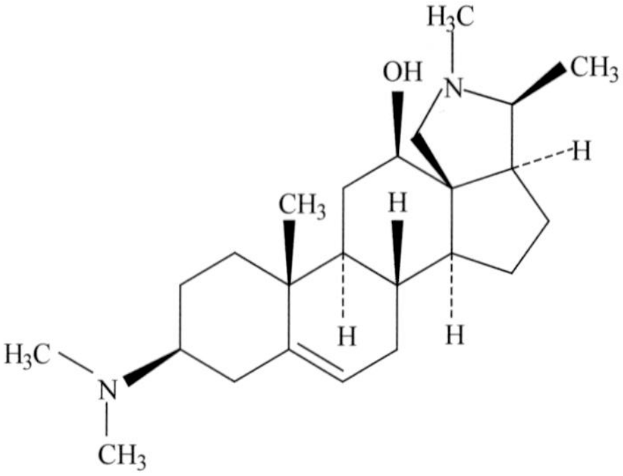

Fig. 245. Structure of holarrhenine.

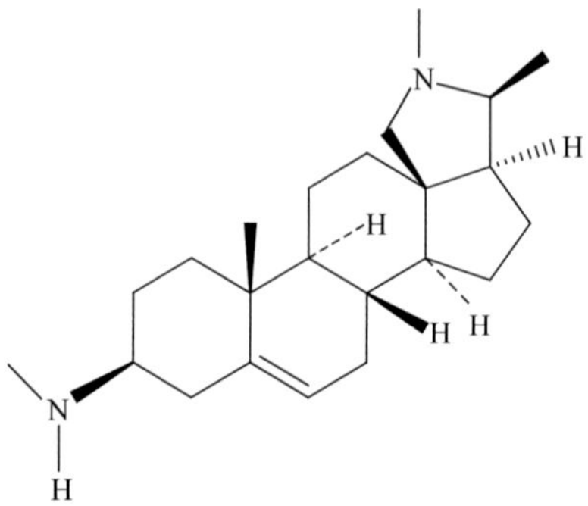

Fig. 246. Structure of kurchine.

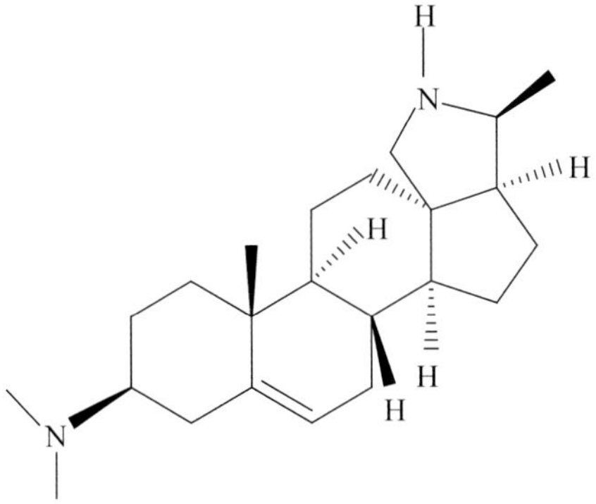

Fig. 247. Structure of conesimine.

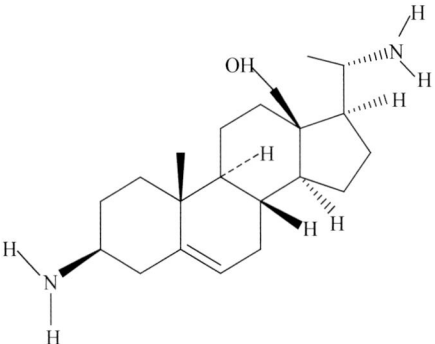

Fig. 248. Structure of holarrhimine.

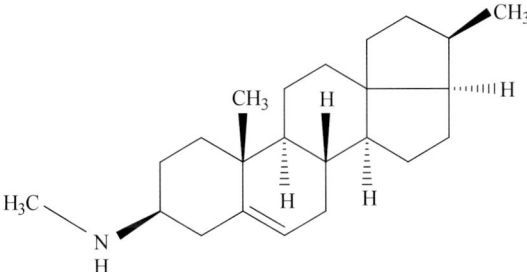

Fig. 249. Structure of norconessine.

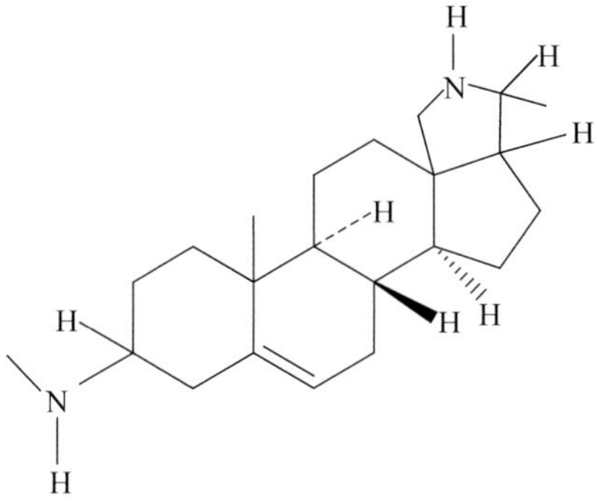

Fig. 250. Structure of conimine.

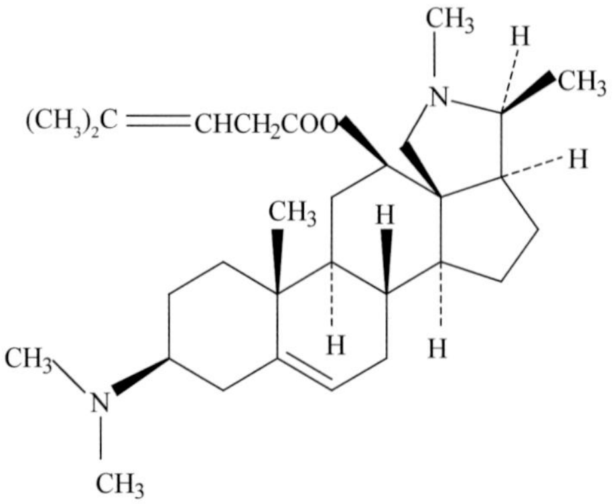

Fig. 251. Structure of isoconessimine.

Therapeutics: It is used in the treatment of amoebiasis, chronic diarrhoea, bleeding piles and fevers.

Parts used: Seeds and bark.

Hyoscyamus albus L.

Family: Solanaceae.

Common name: White henbane.

Distribution: Cultivated in Australia and Japan.

Botany: An herb growing up to 40 or 50 cm. The plant is characterised by woolly light green stems, serrated leaves, calyxes, and fruits. Externally, the flowers are light yellow and dark violet internally. The berries are orange or yellow coloured. The seeds are usually white or gray.

Chemical composition: Tropane alkaloids hyoscyamine, scopolamine, aposcopolamine, norscopolamine, littorine, tropine, cuscohygrine, tigloidine, and tigloyloxytropane.

Actions: Anodyne, antispasmodic, mildly diuretic, hallucinogenic, hypnotic, mydriatic, narcotic and sedative.

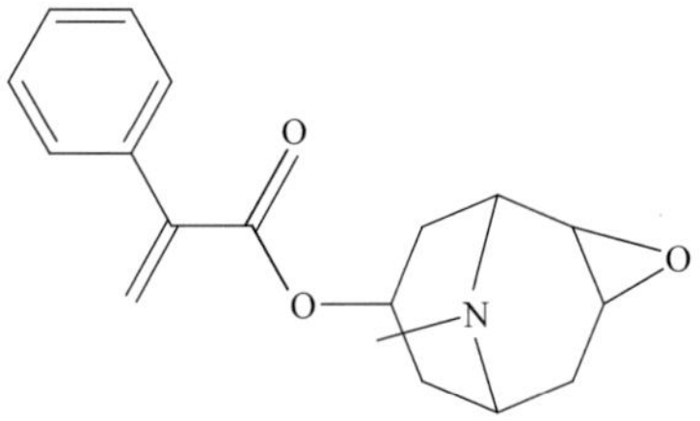

Fig. 252. Chemical structure of aposcopolamine.

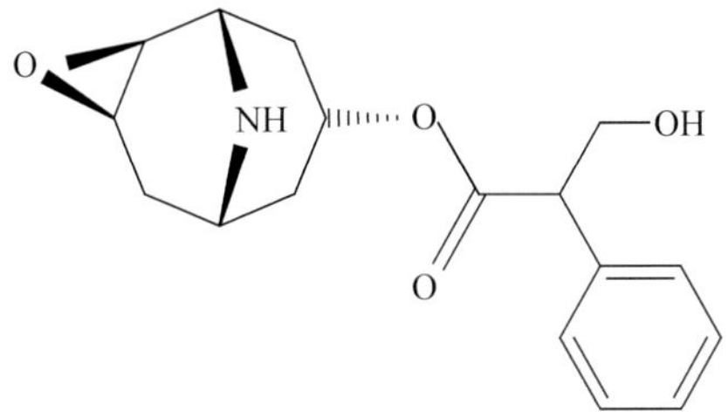

Fig. 253. Chemical structure of norscopolamine.

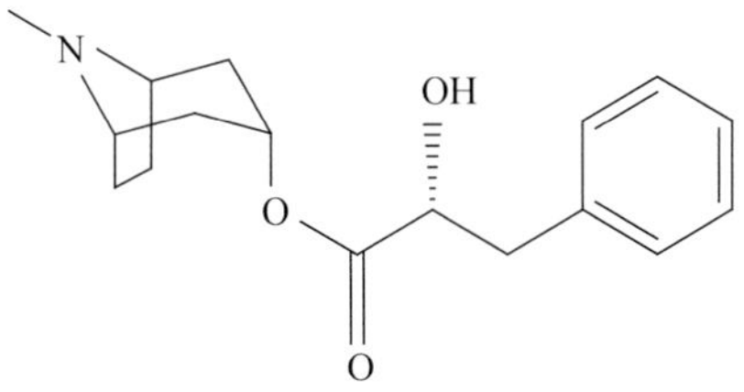

Fig. 254. Chemical structure of littorine.

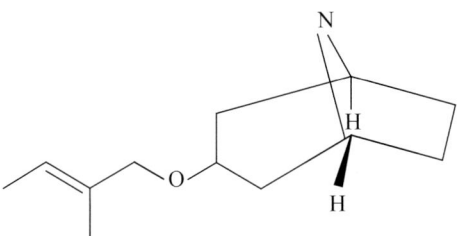

Fig. 255. Chemical structure of tigloyloxytropane.

Therapeutics: Asthma, whooping cough, motion sickness, Meniere's syndrome, tremor in senility or paralysis.

Parts used: Dried leaves.

Hyoscyamus muticus L.

Common name: Egyptian henbane.

Family: Solanaceae.

Distribution: Egypt.

Botany: A succulent perennial herb or shrub growing up to 30–60 cm. The leaves are alternate, long petioled below and succulent. Inflorescence in one-sided spike or raceme-like with dense flowers. Flowers are bisexual. Fruits are small having numerous seeds.

Chemical composition: Tropane alkaloids.

Actions: Anodyne, antispasmodic, mildly diuretic, hallucinogenic, hypnotic, mydriatic, narcotic and sedative.

Therapeutics: Asthma, whooping cough, motion sickness, Meniere's syndrome, tremor in senility or paralysis.

Part used: Dried leaves.

Hyoscyamus niger L.

Common name: Black henbane.

Family: Solanaceae.

Distribution: Europe, Northwestern Africa, Western Asia, Southern Siberia, naturalised in America and Australia.

Botany: An annual or biennial herb. The stem is erect, 30–60 cm high. Leaves alternate, coarsely toothed, grey-green. Flowers in axils of upper leaves, the petals are grey-yellow with violet veins. The fruit is a many-seeded capsule.

Chemical composition: Tropane alkaloids including hyoscine and hyoscyamine.

Actions: Anodyne, antispasmodic, mildly diuretic, hallucinogenic, hypnotic, mydriatic, narcotic and sedative.

Therapeutics: Asthma, whooping cough, motion sickness, Meniere's syndrome, tremor in senility or paralysis.

Part used: Dried leaves.

Juniperus sabina L.

Common name: Savin.

Family: Cupressaceae.

Distribution: Native to the mountains of central and southern Europe and western and central Asia.

Botany: A shrub growing up to 1–4 m. The leaves exist in two distinct forms. The juvenile needle-like leaves and adult scale-leaves. The cones are berry-like, 5–9 mm in diameter, blue-black with a whitish waxy bloom, and contain 1–3 seeds.

Chemical composition: Essential oil (containing pinene and sabinene), resin, bitter principle (juniperine) and organic acids.

Actions: Diuretic.

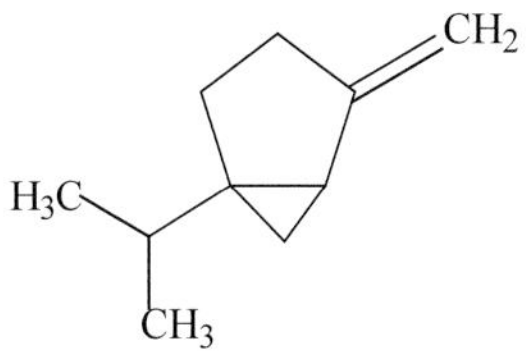

Fig. 256. Structure of sabinene.

Therapeutics: Berries, wood and oil reported to be used in folk remedies for cancer, indurations, polyps, swellings, tumours and warts. The fruit and oil are diuretic, carminative, stimulant, and are used in dropsy, gonorrhea, gleets, leucorrhoea and some cutaneous diseases. The berries are given during scanty urine, cough and pectoral affections. Locally, powder of berries is rubbed on rheumatic and painful swellings. Ash of the bark is applied in certain skin affections. The berries are also recommended during infantile tuberculosis and diabetes.

Part used: Dried leaves.

Lobelia inflata L.

Common name: Indian tobacco.

Family: Campanulaceae.

Distribution: Native to eastern North America.

Botany: An annual or biennial herb. The erect, angular stem, grows up to 1 m. The stem contains a milky sap. The leaves thin, light green in colour, alternate, hairy, ovate, and bluntly serrate. Numerous small, two-lipped, blue flowers grow in spike-like racemes. The fruit is a two-celled capsule having small brown seeds.

Chemical composition: Alkaloids: lobeline, isolobeline, lobelanine, lobelanidine and a bitter glycoside (lobelacrin), a pungent oil (lobelianin) and resin.

Actions: Antispasmodic, entheogenic and emetic.

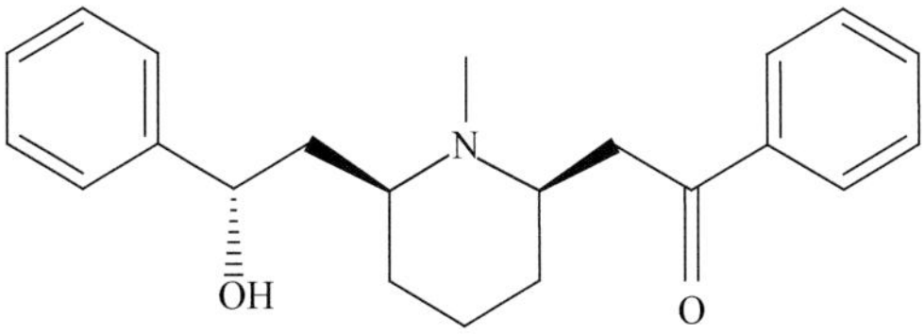

Fig. 257. Chemical structure of lobeline.

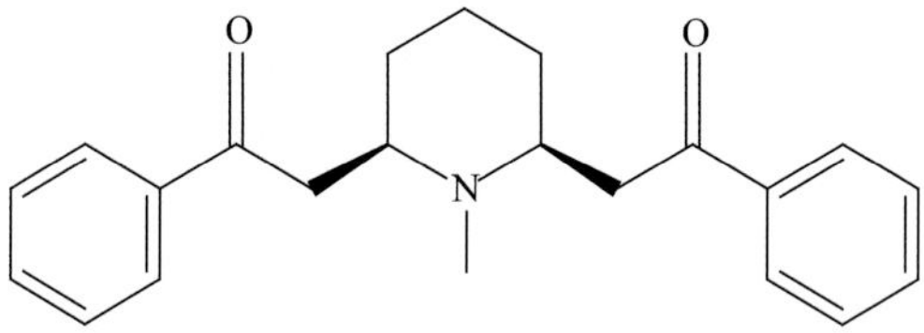

Fig. 258. Chemical structure of lobelanine.

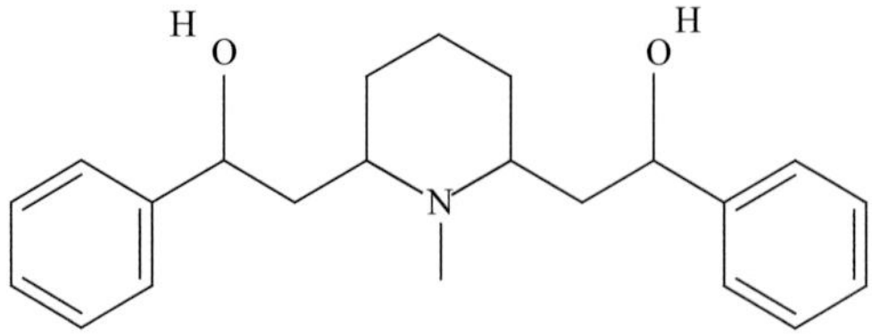

Fig. 259. Chemical structure of lobelanidine.

Therapeutics: Bronchitis and bronchial asthma.

Part used: Dried leaves.

Lobelia nicotianaefolia Roth ex Roem. and Schult.

Common name: Indian tobacco.

Family: Campanulaceae.

Distribution: The Western Ghats and is also found in Deccan and Konkan at altitudes of 900–2,100 m.

Botany: A tall, erect, much branched, hairy herb growing upto 1.5 to 3 meters. The leaves resemble with *Nicotina tabaccum* (tobacco). They are narrowly obovate-lanceolate. Flowers are large, white and borne on terminal compound inflorescences. Seeds are numerous and very small.

Chemical composition: Alkaloids: lobeline and others. The lobeline content of *L. nicotianaefolia* is higher than *L. inflata.*

Actions: Expectorant, emetic, anti-asthmatic, stimulant, antispasmodic, diaphoretic, diuretic, and nervine.

Therapeutics: Pain and snake bites.

Part used: Dried leaves.

Mallotus philippinensis (Lam.) Muell. Arg.

Common name: Kamala tree.

Family: Euphorbiaceae.

Distribution: South East Asia.

Botany: A tree growing up to 10 meters. Leaves are articulated, alternate, rusty-tomentose, ovate or rhombic ovate. Flowers dioecious. Capsule trigonous-globular, having a covering of bright crimson layer of detachable reddish powder.

Chemical composition: Polyphenols: Rottlerin (mallotoxin) and isorottlerin.

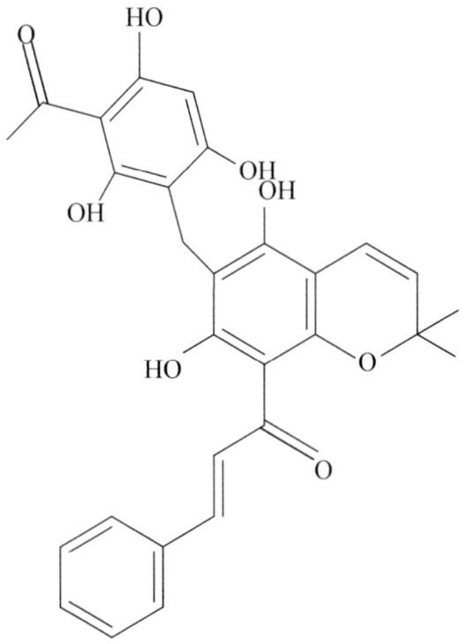

Fig. 260. Chemical structure of rottlerin.

Actions: Anthelmintic.

Therapeutics: Bronchitis, abdominal diseases, and spleen enlargement.

Part used: All parts.

Mandragora autumnalis L.

Common name: Mandrake.

Family: Solanaceae.

Distribution: Mediterranean.

Botany: *M. autumnalis* is a herbaceous perennial, with a large upright tap-root. There is little or no stem, the leaves being borne in a basal rosette upto 60 cm across. The flowers are clustered at the centre of the rosette, each with five sepals, five petals and five stamens. The ovary has two chambers and a long style. The fruit is a fleshy berry with many seeds.

Chemical composition: Tropane alkaloids: hyoscyamine, hyoscine, cuscohygrine, apoatropine, 3-alpha-tigloyloxytropane, 3-alpha,6-beta-ditigloyloxytropane and belladonnine.

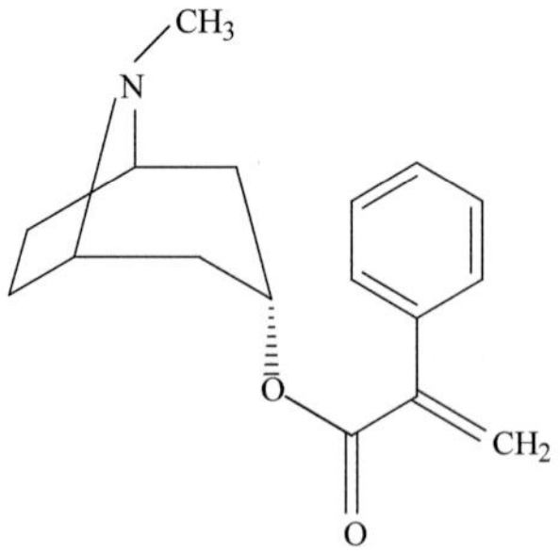

Fig. 261. Chemical structure of apoatropine.

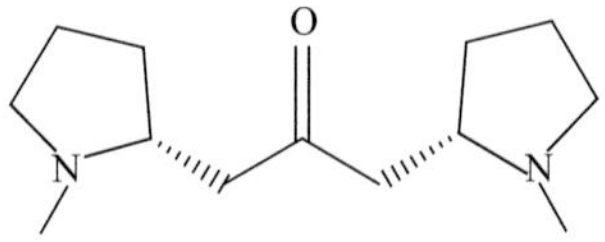

Fig. 262. Chemical structure of cuscohygrine.

Actions: Aphrodisiac, hypnotic, emetic, purgative, sedative and analgesic.

Therapeutics: Insomnia and depression.

Part used: Dried leaves.

Papaver somniferum L.

Common name: Opium poppy.

Family: Papaveraceae.

Distribution: *P. somniferum* is cultivated in India, China, Africa and United Kingdom.

Botany: A small annual herb. The stem and leaves have a covering of coarse hairs. The flowers have four petals having white, mauve or red colour. The fruit is a hairless, rounded capsule. The plant yields opium which is derived from dried latex. The latex is obtained by giving incisions in capsule or fruits.

Chemical composition: Alkaloids (codeine, narceine, narcotine, morphine and papaverine), bitter principle (meconin) and meconic acid.

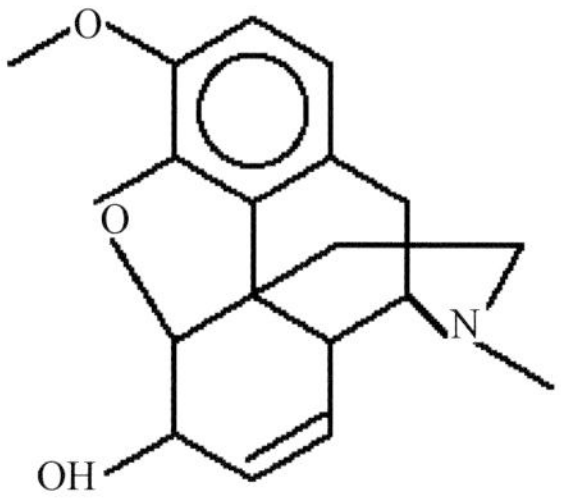

Fig. 263. Chemical structure of codeine.

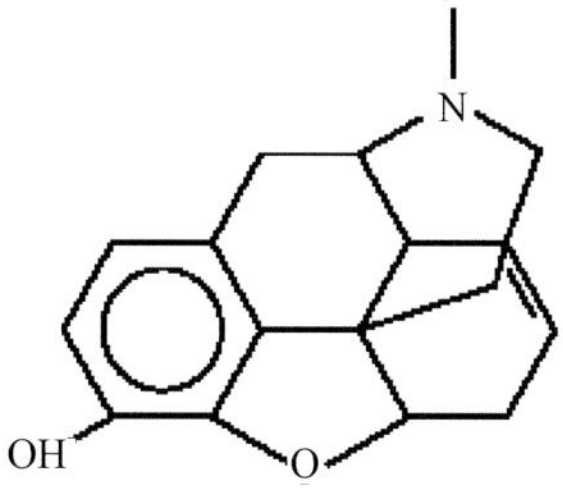

Fig. 264. Chemical structure of morphine.

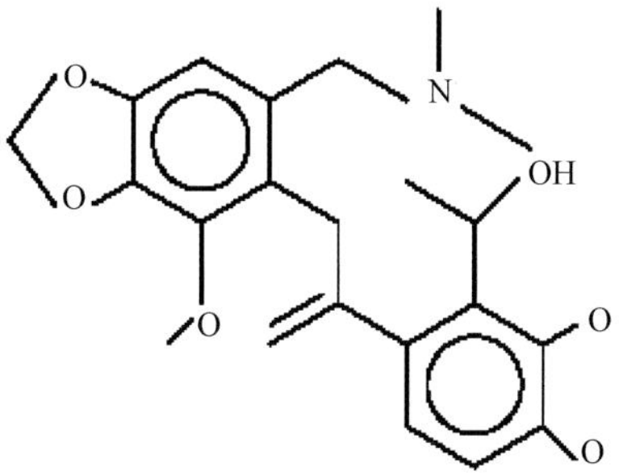

Fig. 265. Chemical structure of narceine.

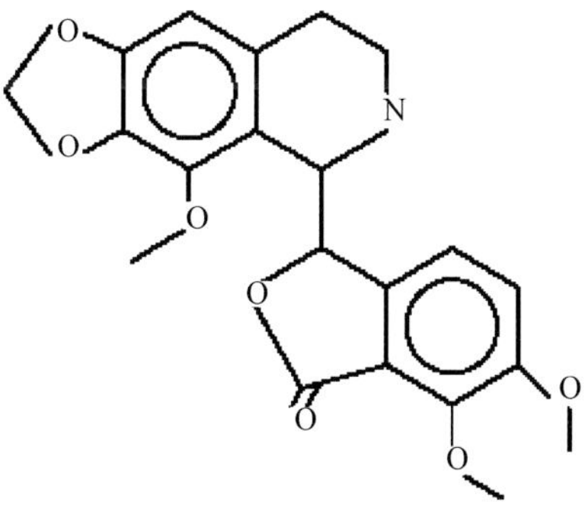

Fig. 266. Chemical structure of narcotine.

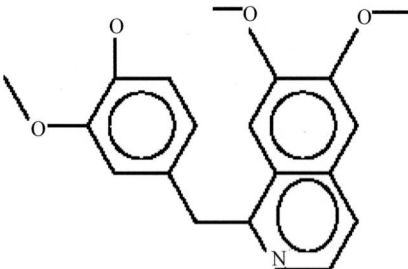

Fig. 267. Chemical structure of papaverine.

Action: Narcotic.

Therapeutics: Used in diarrhoea, cough and internal haemorrhage.

Parts used: Capsule.

Pausinystalia johimbe (K. Schum.) Pierre ex Beille

Common name: Yohimbe.

Family: Rutaceae.

Distribution: West and Central Africa.

Botany: *P. johimbe* grows about 30 m tall. The bark is grey to reddish-brown in colour. It is characterised by longitudinal fissures. The inner bark is pinkish in colour and fibrous in nature. The leaves grow in groups of three and has short stems. The blades are oval-shaped.

Chemical composition: Yohimbine, corynanthine (rauhimbine), and raubasine (ajmalicine) and rauwolscine.

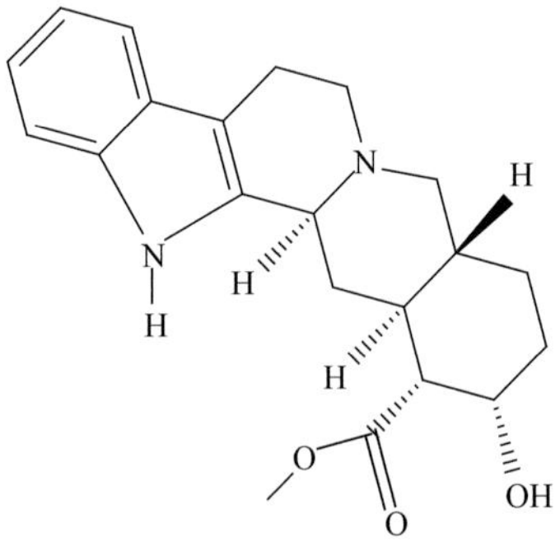

Fig. 268. Chemical structure of yohimbine.

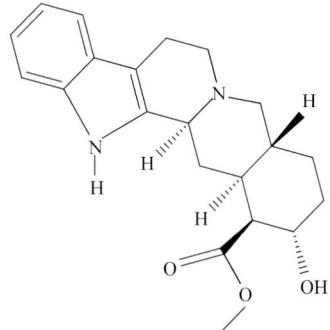

Fig. 269. Chemical structure of corynanthine.

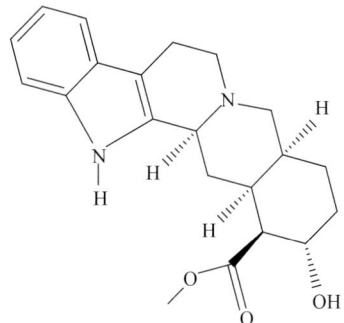

Fig. 270. Chemical structure of rauwolscine.

Actions: Aphrodisiac.

Therapeutics: Erectile dysfunction and impotency.

Part used: Dried leaves.

Toxicity: In larger doses, it induces hypertension.

Pilocarpus jaborandi Holmes.

Common name: Jaborandi.

Family: Rutaceae.

Distribution: Native to northern Brazil.

Botany: *P. jaborandi* is a shrub growing from 4 to 5 feet high. The bark is smooth and greyish. The leaves are large, compound, pinnate with an odd terminal leaflet, with two to four pairs of leaflets. The flowers are thick, small and reddish-purple in colour.

Chemical composition: Alkaloid: pilocarpine, isopilocarpine, pilosine and pilocarpidine.

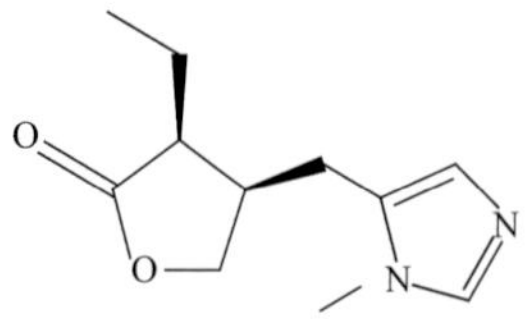

Fig. 271. Chemical structure of pilocarpine.

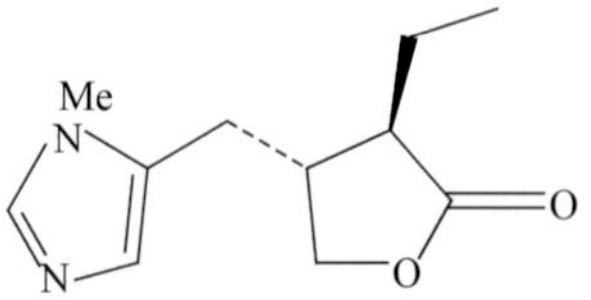

Fig. 272. Chemical structure of isopilocarpine.

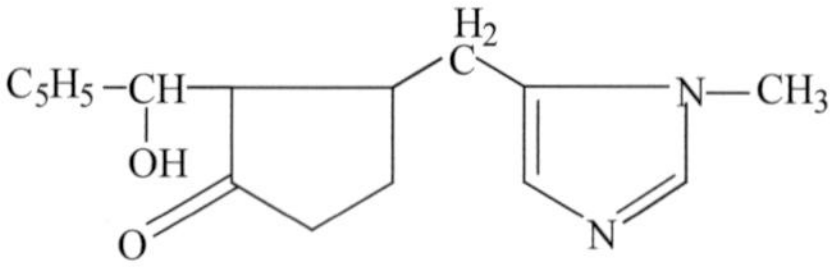

Fig. 273. Chemical structure of pilosine.

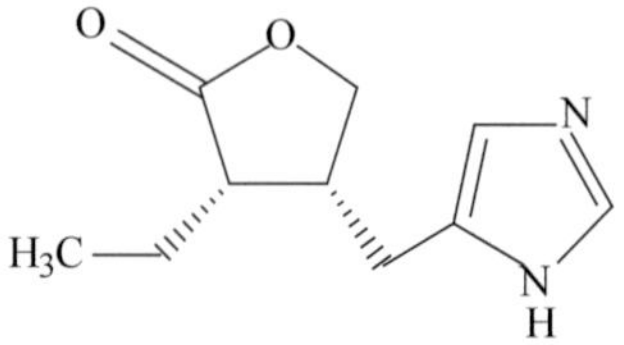

Fig. 274. Chemical structure of pilocarpidine.

Actions: Diaphoretic.

Therapeutics: Diarrhoea and glaucoma. Pilocarpine salts are valuable in ophthalmic practice and are used in eye-drops as miotics and for the treatment of glaucoma. Pilocarpine is a cholinergic agent and stimulates the muscarinic receptors in the eye, causing constriction of the pupil and enhancement of outflow of the aqueous humour.

Part used: Dried leaves.

Pilocarpus microphyllus Stapf ex Wardleworth

Common name: The Maranham Jaborandi.

Family: Rutaceae.

Distribution: Native to northern Brazil.

Botany: A succulent perennial herb or shrub growing over 1 metre. Stem is long and much branched in the upper part. Lower leaves are ovate to rectangular, acute, cuneate or truncate. Calyx is 2–3 cm, having a short triangular blunt teeth. Corolla is 2 × 2 cm, white or green. Capsule is 6 mm.

Chemical composition: Alkaloid: pilocarpine and epiisopiloturine.

Actions: Diaphoretic.

Therapeutics: Diarrhoea and glaucoma.

Part used: Dried leaves.

Piper methysticum G. Forst.

Common name: Kava-kava.

Family: Piperaceae.

Distribution: Western Pacific islands.

Botany: *P. methysticum* is a sparingly branched, erect shrub is very hardy and grows up to 12 feet high. The root is thick, soft wooded when fresh, hardening as it dries.

Chemical composition: Kavalactones: kavain, dihydrokavain, methysticin, dihydromethysticin, yangonin, and desmethoxyyangonin and alkaloid: pipermethystine.

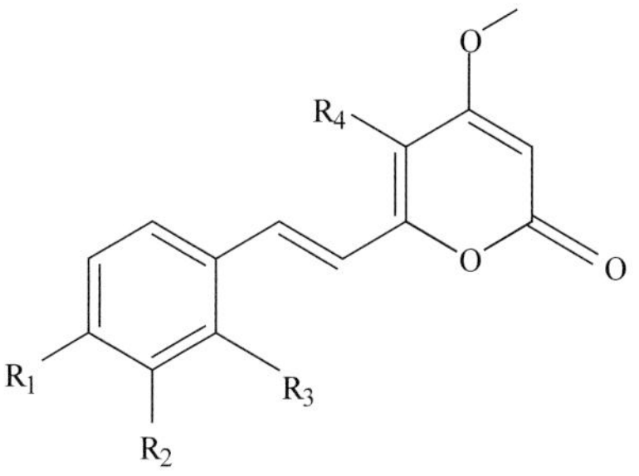

Fig. 275. General chemical structure of kavalactones.

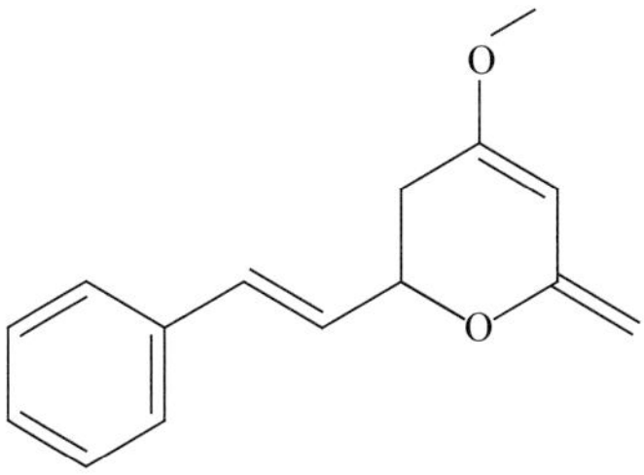

Fig. 276. Chemical structure of kavain.

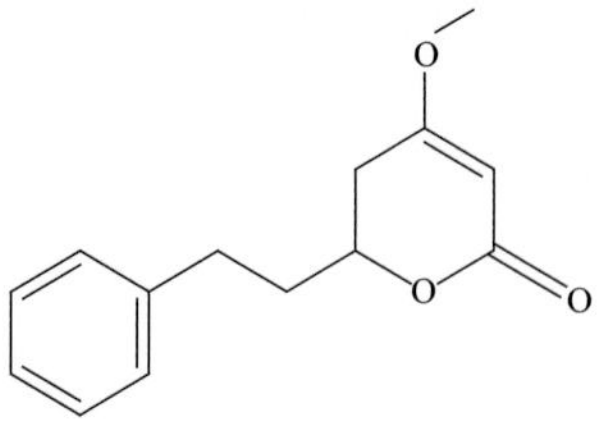

Fig. 277. Chemical structure of dihydrokavain.

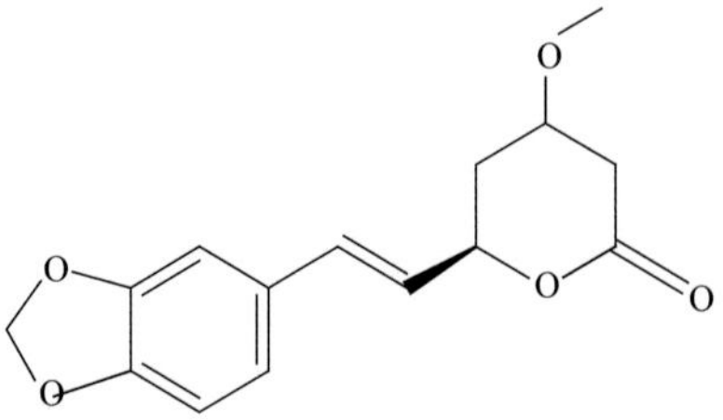

Fig. 278. Chemical structure of methysticin.

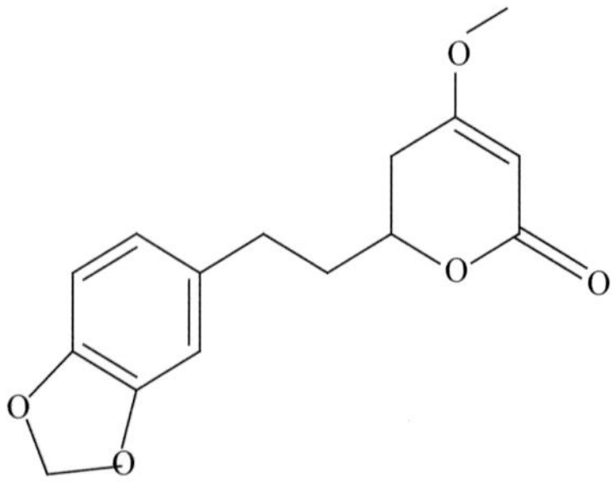

Fig. 279. Chemical structure of dihydromethysticin.

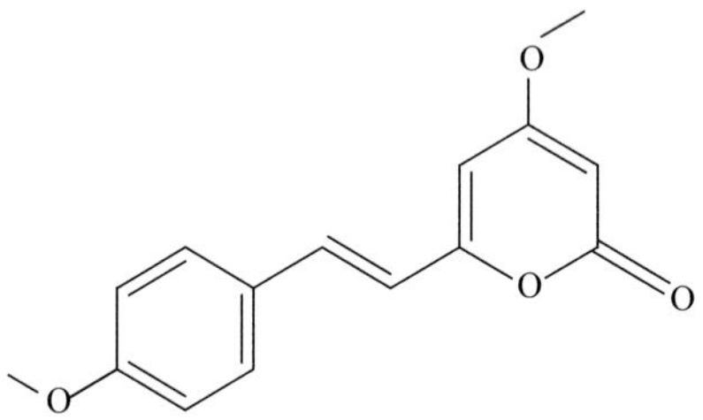

Fig. 280. Chemical structure of yangonin.

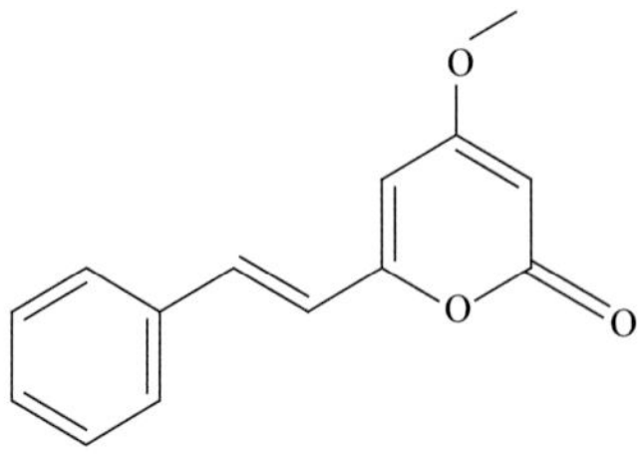

Fig. 281. Chemical structure of desmethoxyyangonin.

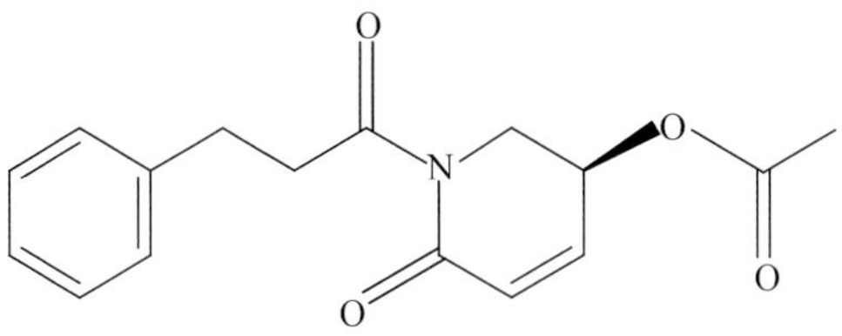

Fig. 282. Chemical structure of pipermethystine.

Actions: Anxiolytic, sedative, anaesthetic, euphoriant, and entheogenic.

Therapeutics: Anxiety neurosis.

Part used: Roots.

Toxicity: As pre-research studies show, methysticin and dihydromethysticin have CYP1A1 inducing effects which account for their toxicity.

Podophyllum hexandrum Royle.

Syn: *P. emodi* Wall.

Common name: Indian podophyllum.

Distribution: *P. hexandrum* is found in The Himalayas, Kashmir and Madagascar.

Botany: *P. hexandrum* is a small herb having simple branch system. Leaves are dissected and oval. Flowers are white or pink. Fruit is egg shaped and seeds are many. Root is perennial, brown and having distinct smell. The aerial part of the plant gets destroyed in winter months.

Chemical composition: Resin (podophyllotoxin) and glucoside. Podophyllin is light brown to green-yellow coloured powder with typical herbaceous odour and bitter taste. It is soluble in ether and alcohol but slightly soluble in water.

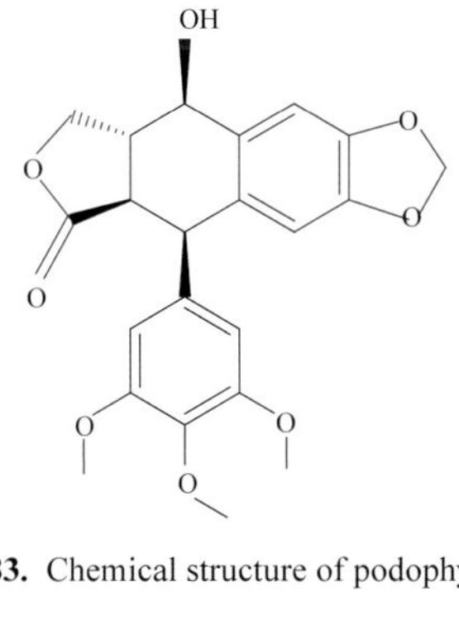

Fig. 283. Chemical structure of podophyllotoxin.

Action: Anti-cancer.

Therapeutics: *P. hexandrum* is used in the treatment of chronic constipation, diabetes mellitus, hypertension and cancer. The paint of the drug is applied on the warts.

Parts used: Roots.

Rauvolfia serpentina (L.) Benth. ex Kurz

Synonyms: *Ophioxylon album* Gaertn., *Ophioxylon obversum* Miq., *Ophioxylon salutiferum* Salisb., *Ophioxylon serpentinum* L., *Ophioxylon trifoliatum* Gaertn., *Rauvolfia obversa* (Miq.) Baill. and *Rauvolfia trifoliata* (Gaertn.) Baill.

Common name: Indian snakeroot.

Family: Apocynaceae.

Distribution: Native to the Indian subcontinent and East Asia.

Botany: *R. serpentina* is an evergreen shrub. The leaves have acute apex; upper leaves are light green in colour and lower leaves are dark green in colour. Roots are broken, have soft bark and are brown in colour. Flowers are pinkish, white or purple. The fruit resembles with pea and turns black when gets ripe.

Chemical composition: Alkaloids: ajmaline, ajmalicine, desperidine and reserpine.

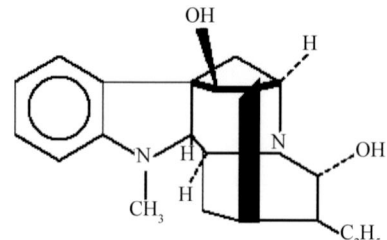

Fig. 284. Chemical structure of ajmaline.

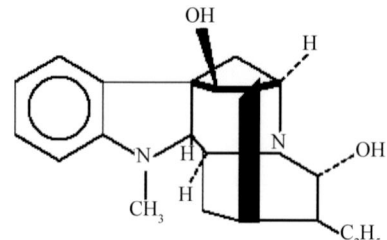

Fig. 285. Chemical structure of ajmalicine.

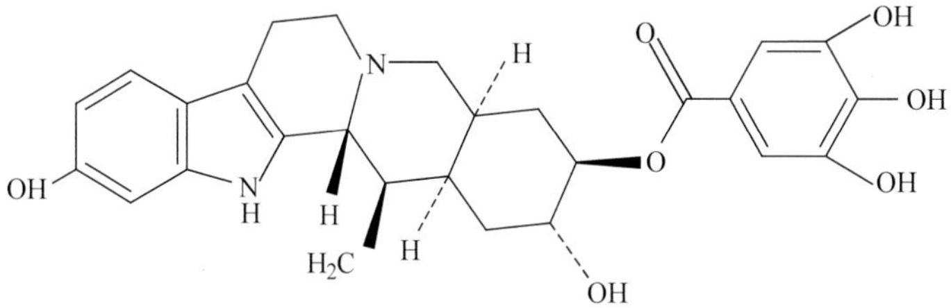

Fig. 286. Chemical structure of desperidine.

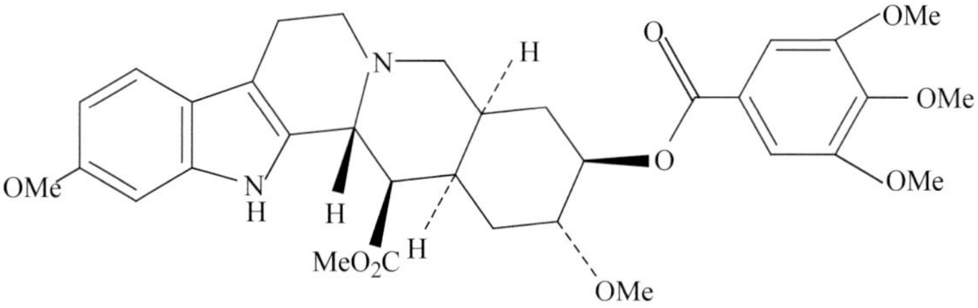

Fig. 287. Chemical structure of reserpine.

Actions: Hypnotic.

Therapeutics: Hypertension, insanity and insomnia. The juice of the leaves is used for reducing opacity of cornea.

Part used: Root bark.

Rauvolfia vomitoria Afzel.

Synonyms: *Hylacium owariense* Afzel. *Rauvolfia senegambiae* A. DC.

Common name: African serpentwood, Poison devil's-pepper.

Family: Apocynaceae.

Distribution: Native to tropical Africa.

Botany: A shrub or small tree growing up to 8 m tall. The leaves grow in whorls of three and are elliptic and pointed at the end. Flowers are tiny, sweet-scented, pale greenish-white in colour. The fruits are orange in colour containing a single seed.

Chemical composition: Alkaloids: resperine and alstonine and benzoquinone: 2,6-Dimethoxybenzoquinone.

Actions: Emetic and sedative.

Therapeutics: *R. vomitoria* is used by Nigerian traditional healers to treat psychiatric patients. In the Democratic Republic of Congo, *R. vomitoria* is used to treat leprosy.

Part used: Root bark.

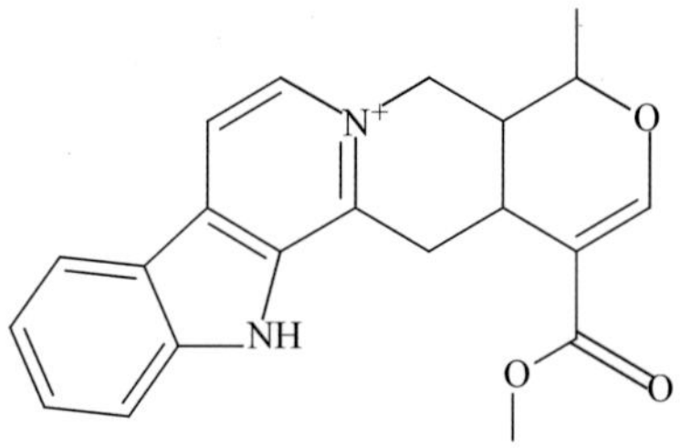

Fig. 288. Chemical structure of alstonine.

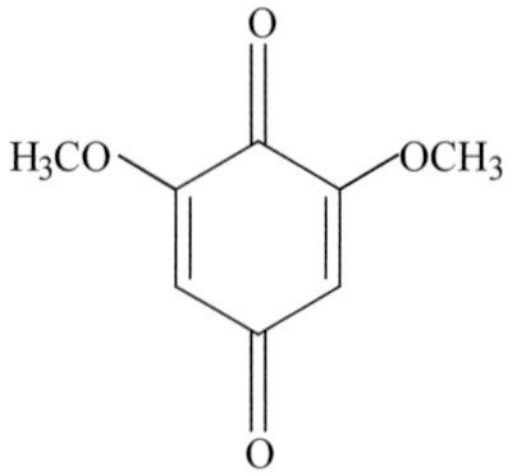

Fig. 289. Chemical structure of 2,6-Dimethoxybenzoquinone.

Rhus radicans L.

Syn: *Toxicodendron radicans* (L.) Kuntze

Common name: Poison Ivy or eastern poison ivy.

Family: Anacardiaceae.

Distribution: Throughout America.

Botany: *R. radicans* is grown as vine or shrub. The alternate compound leaves of three leaflets is the distinguishing character. The new leaves are shiny. Leaf margin can be smooth, toothed and lobed. The color of the leaf changes with age and season.

Chemical composition: Resin: urushiol.

Toxicity: It can cause contact dermatitis, an allergic reaction due to urushiol.

Part used: Dried leaves.

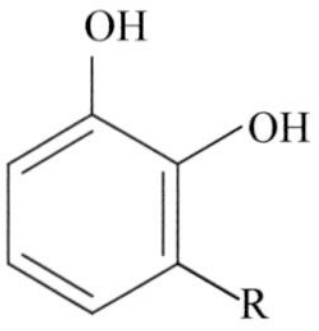

$R = (CH_2)_{14}CH_3$

Fig. 290. Chemical structure of urushiol.

Schoenocaulon officinale (Schltdl. and Cham.) A.Gray ex Benth.

Common name: Sabadilla, Cevadilla.

Family: Rutaceae.

Distribution: South America.

Botany: *S. officinale* grows up to 3–4 feet high. The leaves are numerous, spreading on the ground, all radical, ovate-oblong, and obtuse. The flowers are blackish-purple in colour. Seeds 3 in each cell.

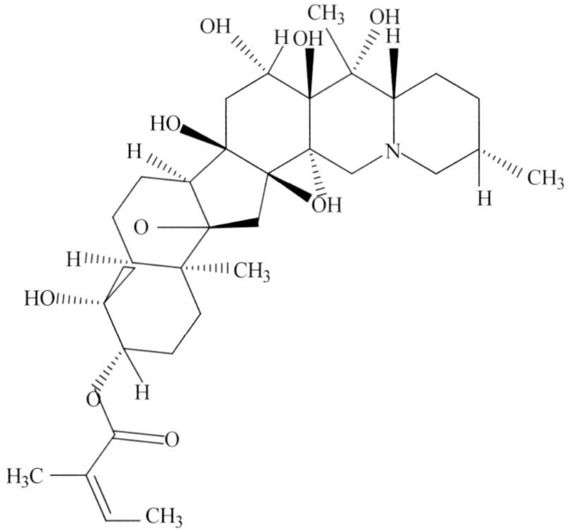

Fig. 291. Chemical structure of veratrine.

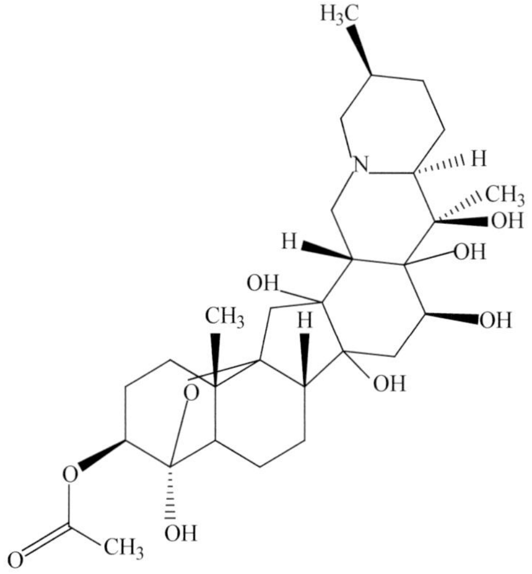

Fig. 292. Chemical structure of cevacine.

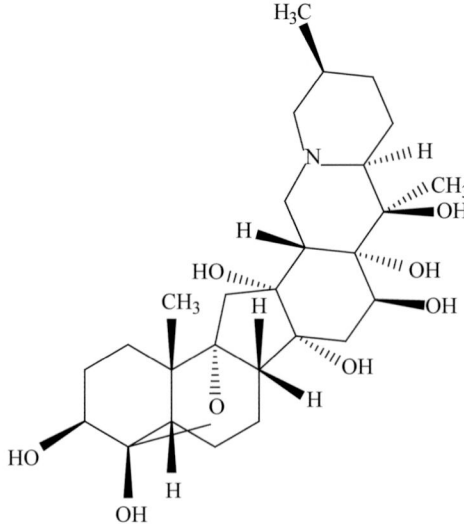

Fig. 293. Chemical structure of protocevine.

Chemical composition: Alkaloids: veratrine, cevacine, protocevine and vanilloylcevine.

Actions: Vermifuge.

Therapeutics: Tapeworm infestation in the nervous system.

Part used: Seeds.

Scopolia carniolica Jacq.

Common name: European scopolia.

Family: Solanaceae.

Distribution: South-eastern Europe.

Botany: *S. carniolica* grows to the height of 1 foot, and has thin leaves. Its fruit is transversely dehiscent capsule. The rhizome is horizontal, curved, almost cylindrical, and somewhat flattened vertically..

Chemical composition: Tropane alkaloid: atropine and scopolamine.

Actions: Narcotic and mydriatic.

Therapeutics: Mania, hysteria, and drug addictions.

Part used: Dried rhizome.

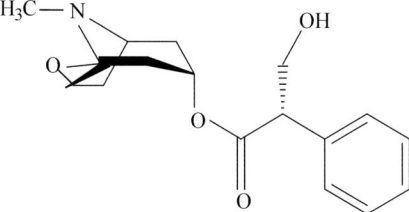

Fig. 294. Chemical structure of scopolamine.

Scopolia japonica Max.

Common name: Japanese belladonna.

Family: Solanaceae.

Distribution: South-Eastern Europe.

Botany: A perennial herb.

Chemical composition: Coumarins: umbelliferone and scopoletin.

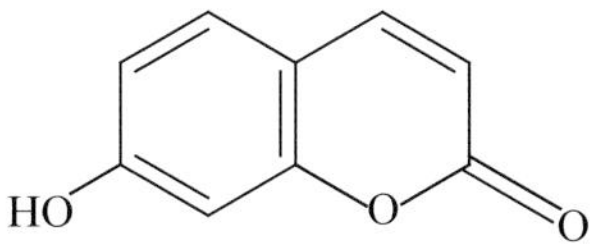

Fig. 295. Chemical structure of umbelliferone.

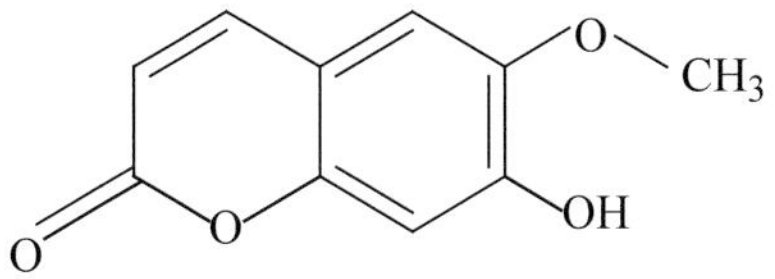

Fig. 296. Chemical structure of scopoletin.

Actions: Narcotic and mydriatic.

Therapeutics: Mania, hysteria, and drug addictions.

Part used: Dried rhizome.

Stephania tetrandra (Willd) Walp.

Family: Menispermaceae.

Distribution: Native to China and Taiwan.

Botany: *S. tetrandra* grows from a short, woody caudex, climbing to a height of around three meters. The leaves are arranged spirally on the stem, and are peltate, i.e., with leaf petiole attached near the centre of the leaf.

Chemical composition: Alkaloids: tetrandrine, fangchinoline, cyclanoline and demethyltetrandrine.

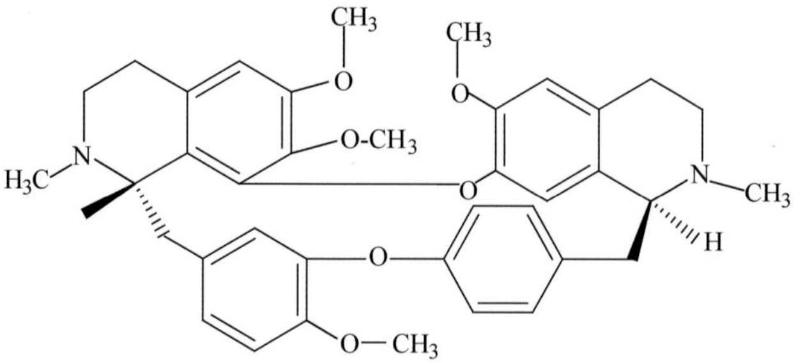

Fig. 297. Chemical structure of tetrandrine.

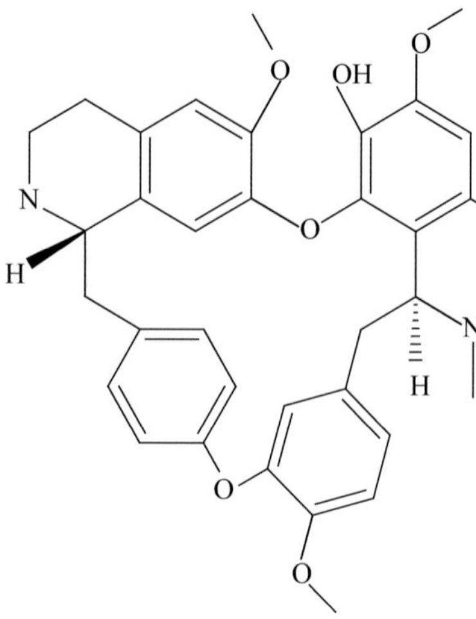

Fig. 298. Chemical structure of fangchinoline.

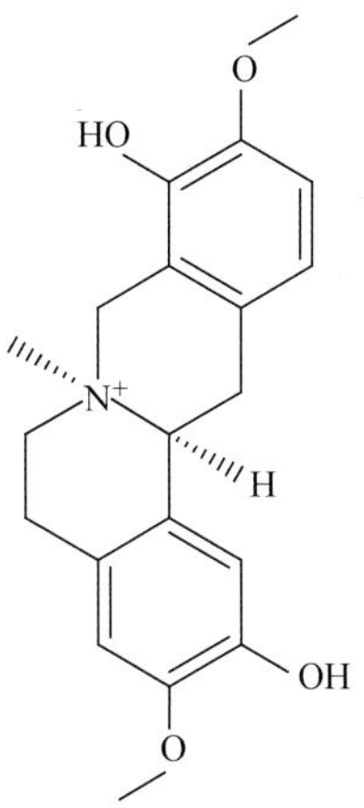

Fig. 299. Chemical structure of cyclanoline.

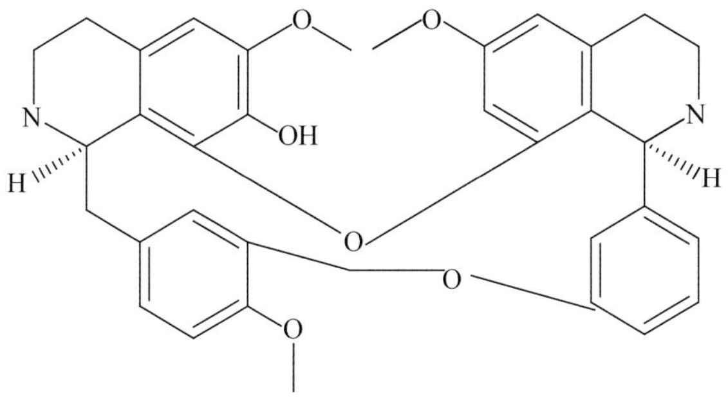

Fig. 300. Chemical structure of demethyltetrandrine.

Actions: In traditional Chinese medicine, *S. tetrandra* is used as analgesic and diuretic.

Therapeutics: Pain.

Part used: Roots.

Stropanthus kombe Oliver

Common name: The kombe arrow poison.

Family: Apocynaceae.

Distribution: Tropical regions of Eastern Africa.

Botany: A deciduous vine growing up to 3.5 meters. The bark is a reddish brown in colour having lenticels with black or dark-brown colour. The roots have thick and fleshy character. The papery leaves are simple, and found in opposite arrangement. 1–12 cream colored flowers can be found on the peduncle. The cyme inflorescence is on the terminal end of the vine.

Chemical composition: Cardiac glycoside: k-strophanthin.

Actions: Cardiac tonic.

Therapeutics: Caridac oedema and anasarca.

Part used: Dried, ripe seeds, deprived of their awns.

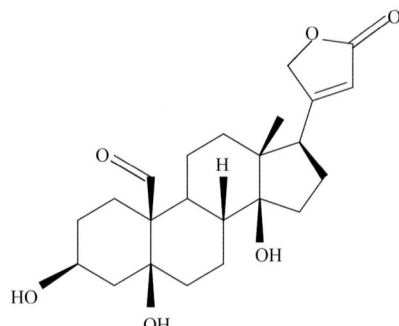

Fig. 301. Structure of k-strophanthin.

Stropanthus courmonti Sacleux ex Franch.

Family: Apocynaceae.

Distribution: Native to Kenya, Tanzania, Malawi, Mozambique, Zambia and Zimbabwe.

Botany: A deciduous plant growing up to 22 metres long. As a shrub it grows up to 4 metres. The fragrant flowers are characterized by a white turning red and purple corolla tube, yellow with purple streaks inside.

Chemical composition: Cardiac glycosides.

Actions: Aphrodisiac.

Therapeutics: Rheumatism.

Part used: Dried, ripe seeds that are deprived of their awns.

Stropanthus nicholsoni E. M. Holmes.

Syn: *S. asper* Oliv. ex L. Planch.

Family: Apocynaceae.

Distribution: Native to Malawi, Mozambique, Zambia and Zimbabwe.

Botany: A deciduous scrambling shrub. Its fragrant flowers feature corolla lobes ending in tails up to 10 centimetres long.

Chemical composition: Cardiac glycosides.

Actions: Aphrodisiac.

Therapeutics: Rheumatism.

Part used: Dried, ripe seeds that are deprived of their awns.

Stropanthus gratus (Wall. and Hook.) Baill.

Common name: Climbing Oleander.

Family: Apocynaceae.

Distribution: Native to West and Central Africa.

Botany: An evergreen vine or shrub growing up to 25 feet or more. The plant has shiny leathery olive green, oblong to elliptical leaves. Flowers are arranged in clusters and are showy, scented and pink or white in colour. Dry fruit contains glabrous seeds.

Chemical composition: Cardiac glycoside: g-strophanthin.

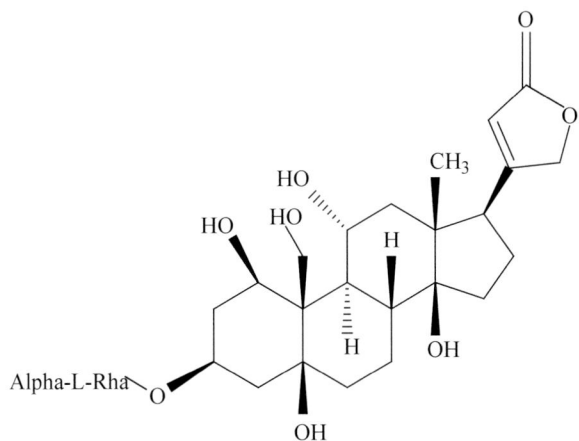

Fig. 302. Structure of g-strophanthin (ouabain).

Actions: Cardiac tonic.

Therapeutics: Cardiac anasarca.

Part used: Dried, ripe seeds, deprived of their awns.

Stropanthus emini Aschers et Pax

Family: Apocynaceae.

Distribution: Native to Malawi, Mozambique, Zambia and Zimbabwe.

Botany: A deciduous shrub.

Chemical composition: Cardiac glycoside: e-strophanthin.

Actions: Cardiac tonic.

Therapeutics: Cardiac oedema.

Part used: Dried, ripe seeds that are deprived of their awns.

Stropanthus sarmentosus DC.

Common name: Spider tresses and poison arrow vine.

Family: Apocynaceae.

Distribution: Native to several African countries.

Botany: A deciduous shrub or as a liana up to 40 metres long, with the stem's diameter up to 15 centimetres. Its fragrant flowers feature a white to purple corolla which is red or purple-streaked on the inside.

Chemical composition: Cardiac glycosides: sarmentoside A, tholloside, sarhamnoloside, locundioside and bipindoside.

Actions: Cardiac tonic.

Therapeutics: Cardiac oedema.

Part used: Dried, ripe seeds that are deprived of their awns.

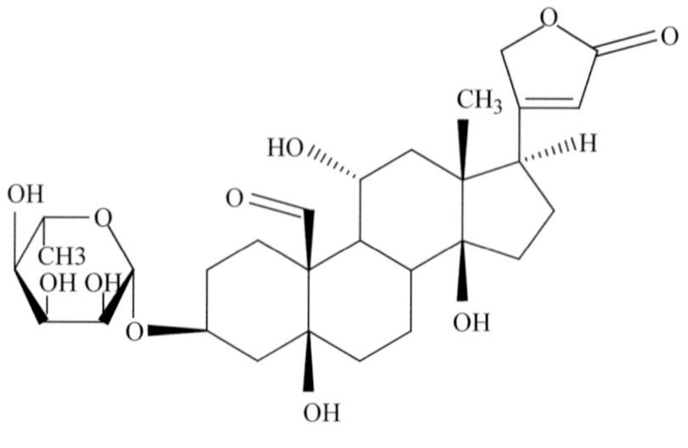

Fig. 303. Chemical structure of sarmentoside A.

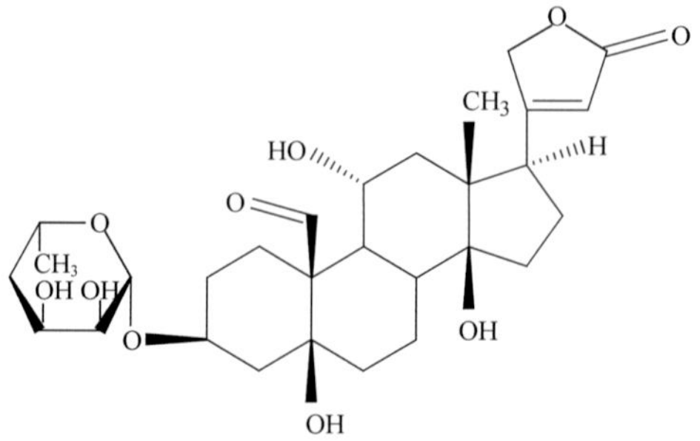

Fig. 304. Chemical structure of tholloside.

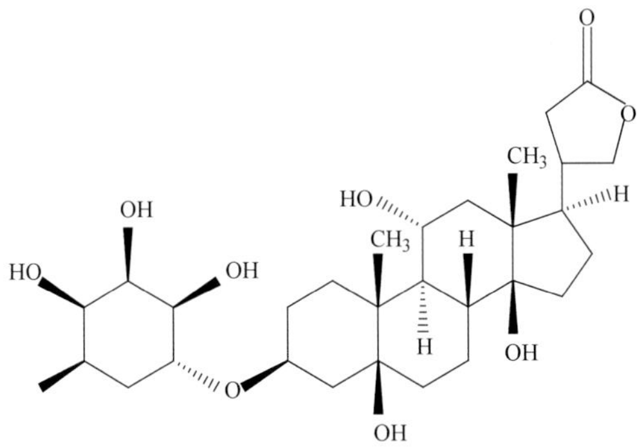

Fig. 305. Chemical structure of locundioside.

Stropanthus hispidus DC.

Common name: The hispid strophanthus.

Family: Apocynaceae.

Distribution: Native to African countries.

Botany: A shrub that can grow up to 5 metres tall. Its flowers feature a yellow corolla. The yellow corona lobes are also spotted with red, purple or brown colour.

Chemical composition: Cardiac glycoside: h-strophanthin.

Actions: Cardiac tonic.

Therapeutics: Cardiac oedema.

Part used: Dried, ripe seeds, deprived of their awns.

Strychnos ignati Lindl.

Common name: Ignatia.

Family: Loganiaceae.

Distribution: Native to the Philippines.

Botany: A large, woody vine. Leaves are opposite, smooth, leathery, oblong or elliptic. Flowers are white, borne mostly in the axils of the upper leaves. Fruit is rounded, pale yellowish and brown, 10 centimeters or more in diameter, containing several seeds.

Chemical composition: Alkaloids: strychnine and brucine.

Actions: Convulsant.

Therapeutics: *S. ignati* is used in the treatment of loss of appetite, nervine disorders, facial palsy, neuralgia, atony of the urinary bladder and general debility. It is mainly used in Homeopathic system of medicine.

Part used: Seeds.

Strychnos nux vomica L.

Common name: Nux vomica.

Family: Loganiaceae.

Distribution: India and Sri Lanka.

Botany: A medium-sized tree with a short thick trunk. The wood is dense, hard white, and close-grained. The branches are covered with a smooth ashen bark. The leaves have an opposite decussate arrangement, short stalked, are oval shaped, also have a shiny coat and are smooth on both sides. The flowers are small with a pale green colour with a funnel shape.

Chemical composition: Alkaloids (strychnine, brucine, icajine and vomicine), glucoside (loganin) and colouring matter.

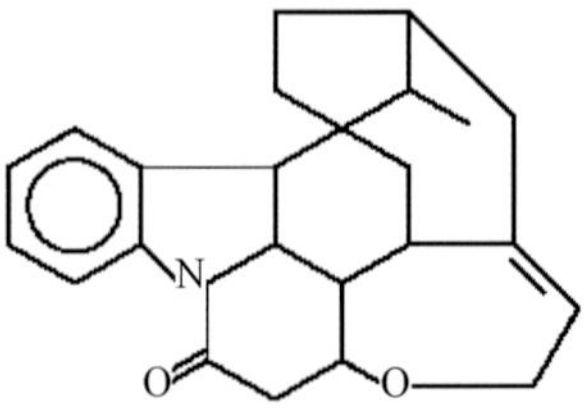

Fig. 306. Structure of strychnine.

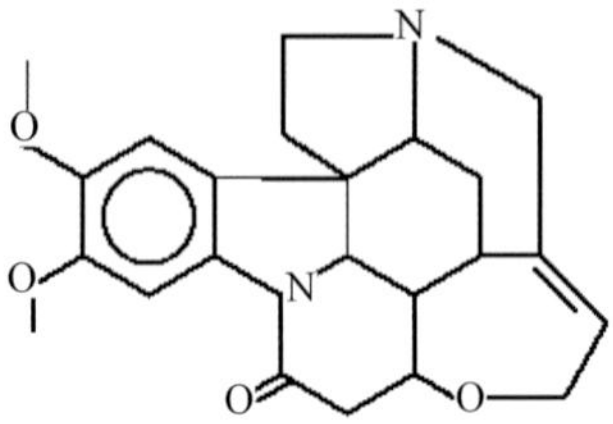

Fig. 307. Structure of brucine.

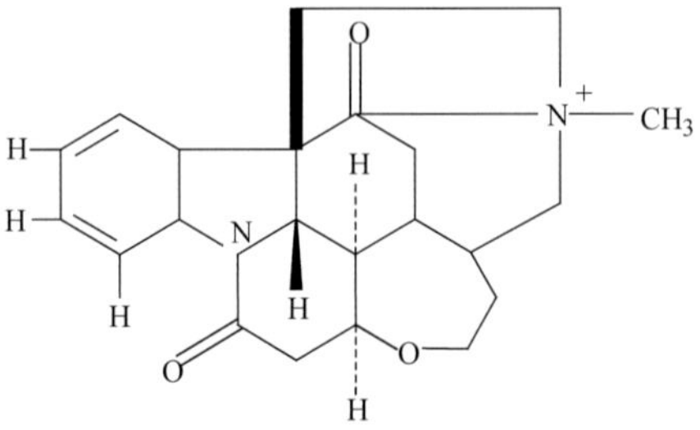

Fig. 308. Structure of Icajine.

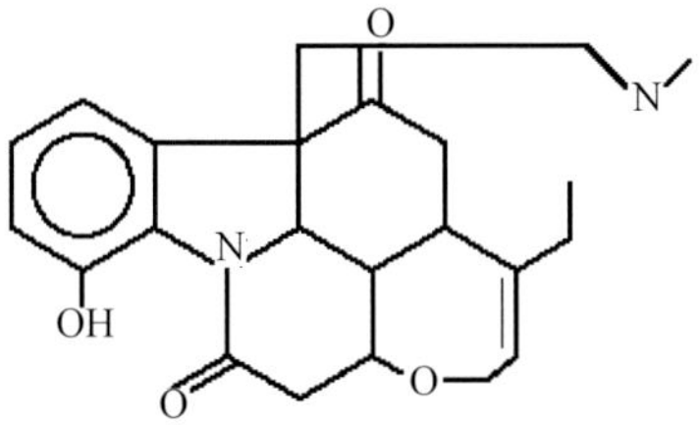

Fig. 309. Structure of vomicine.

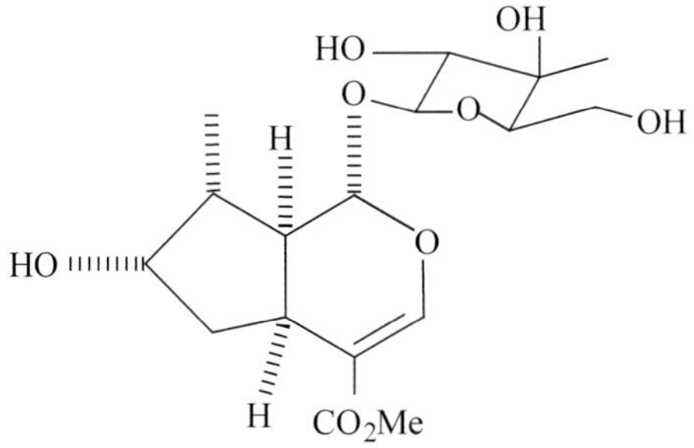

Fig. 310. Structure of loganin.

Actions: Convulsant.

Therapeutics: *S. nuxvomica* is used in the treatment of loss of appetite, nervine disorders, facial palsy, neuralgia, atony of the urinary bladder and general debility.

Part used: Seeds.

Toxicity: The seeds of *S. nuxvomica* if not purified, can be toxic. If given in excess, it acts as convulsant and produces symptoms similar to tetanus. Death occurs due to respiratory failure. It is on the Commission E list of unapproved herbs because it is not recommended for use and has not been proven to be safe or effective.

Ulmus rubra Muhl.

Syn: *Ulmus fulva* Michx., Loudon, Bentley and Trimen, Sarg.

Common name: The slippery elm.

Family: Ulmaceae.

Distribution: Native to eastern North America.

Botany: The trunk is reddish-brown with grey-white bark on the branches. The bark is rough, with vertical ridges. The slippery elm can grow up to 18–20 meters in height. In the spring, dark brown floral buds appear and open into small, clustered flowers at the branch tips.

Chemical composition: Mucilage, phytosterols, sesquiterpenes, calcium oxalate, cholesterol, and tannins.

Actions: Demulcent.

Therapeutics: Gout, rheumatism, cold sores, wounds, abscesses, ulcers, and toothaches.

Part used: Bark.

Note: The Food and Drug Administration has declared *U. rubra* to be a safe and effective oral demulcent.

Veratrum viride Ait.

Common name: Indian hellebore.

Family: Melanthiaceae.

Distribution: Native to eastern and western North America.

Botany: A perennial herb growing up to 0.7–2 metres. The stem is solid green. The leaves are spirally arranged, elliptic to broad lanceolate ending in a short point. The leaves are heavily ribbed and hairy on the underside. The flowers are numerous, produced in a large branched inflorescence 30–70 cm. The fruit is a capsule with flat 8–10 mm diameter seeds.

Chemical composition: Steroidal alkaloids (jervine, pseudojervine, rubijervine, cevadine, germitrine, germidine, veratralbine, and veratroidine) and chelidonic acid.

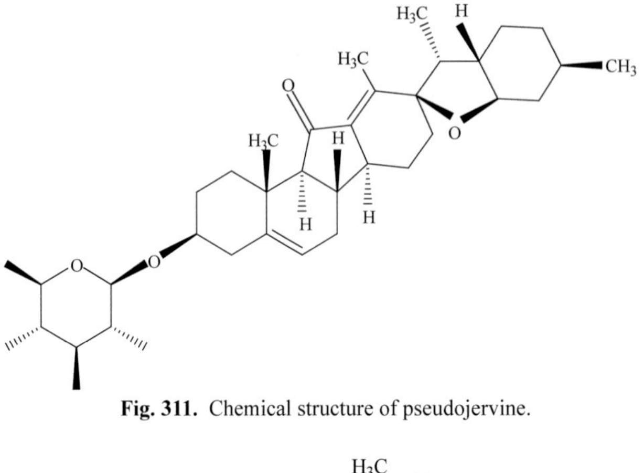

Fig. 311. Chemical structure of pseudojervine.

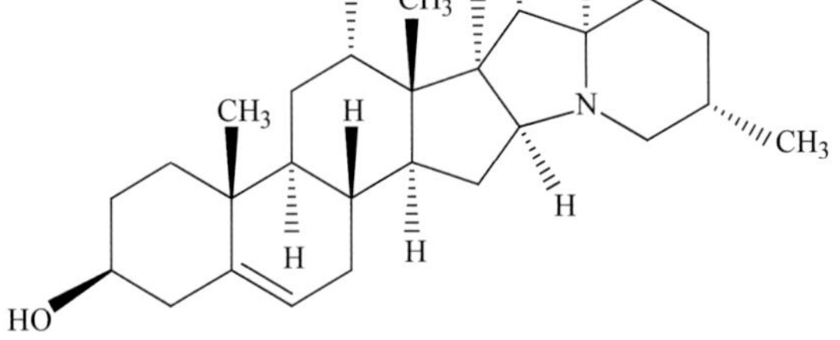

Fig. 312. Chemical structure of rubijervine.

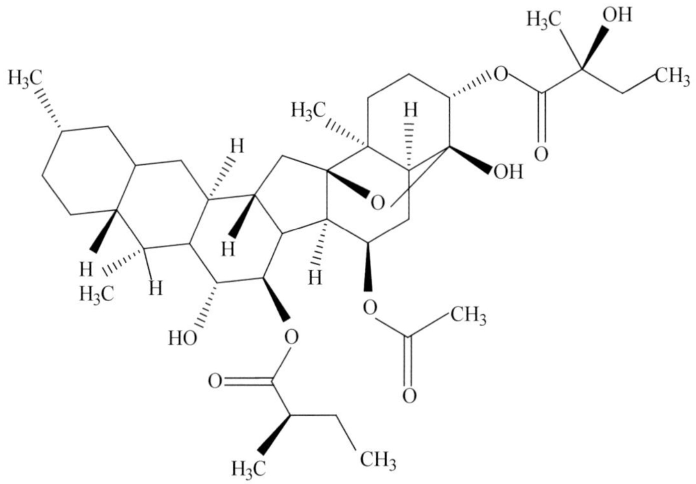

Fig. 313. Chemical structure of germitrine.

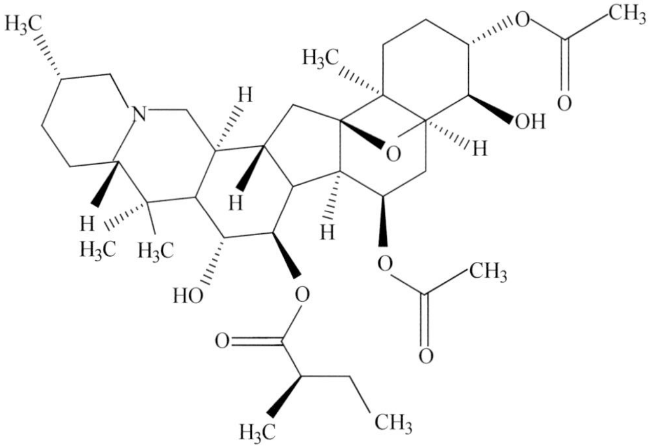

Fig. 314. Chemical structure of germidine.

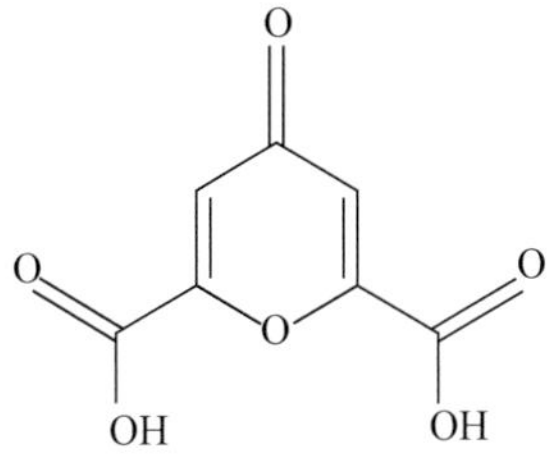

Fig. 315. Chemical structure of chelidonic acid.

Actions: Antihypertensive.

Therapeutics: Hypertension.

Toxicity: The plant is highly toxic.

Part used: Fresh root.

Veratrum album L.

Common name: White hellebore.

Family: Melanthiaceae.

Distribution: Native to Europe and parts of western Asia.

Botany: A perennial herb with a stout vertical rhizome covered with remnants of old leaf sheaths. The stout, simple stems are 50 to 175 cm tall.

Chemical composition: Alkaloids: jervine, veratridine, protoveratrine (A&B), veratramine and cevadine.

Actions: Antihypertensive.

Therapeutics: Hypertension.

Toxicity: The plant is highly toxic.

Part used: Fresh root.

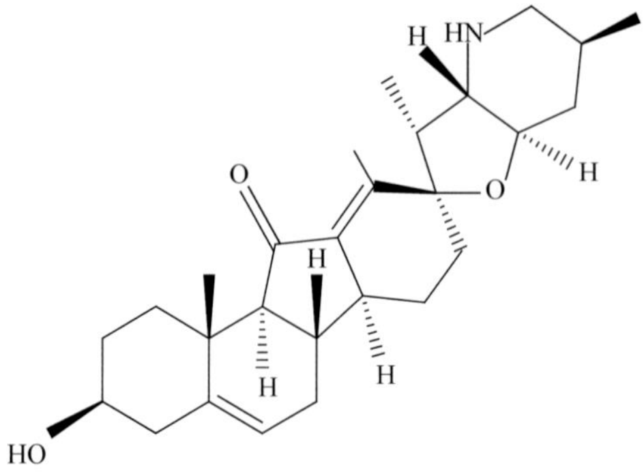

Fig. 316. Structure of jervine.

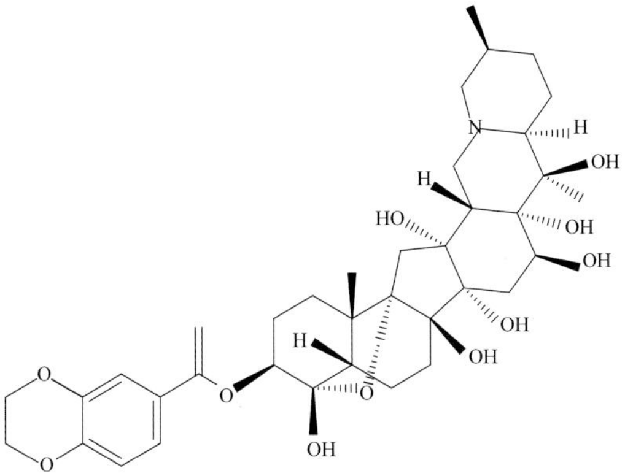

Fig. 317. Structure of veratridine.

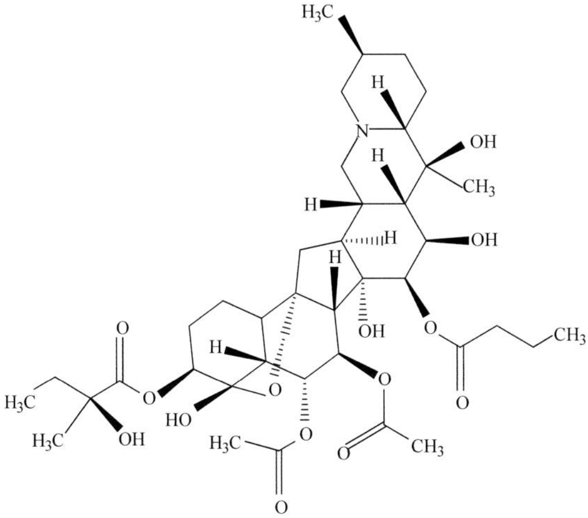

Fig. 318. Structure of protoveratrine A.

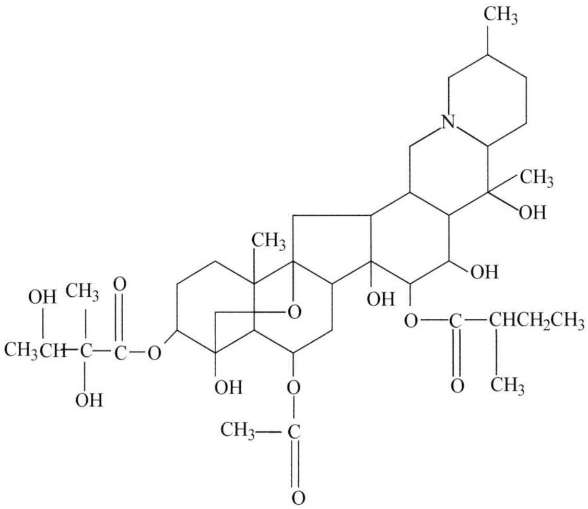

Fig. 319. Structure of protoveratrine B.

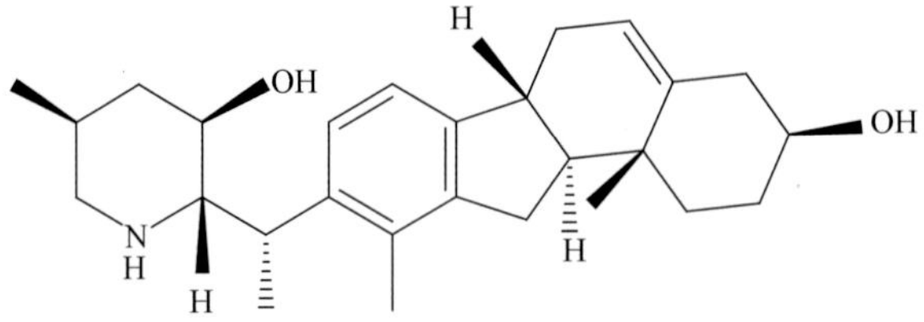

Fig. 320. Structure of veratramine.

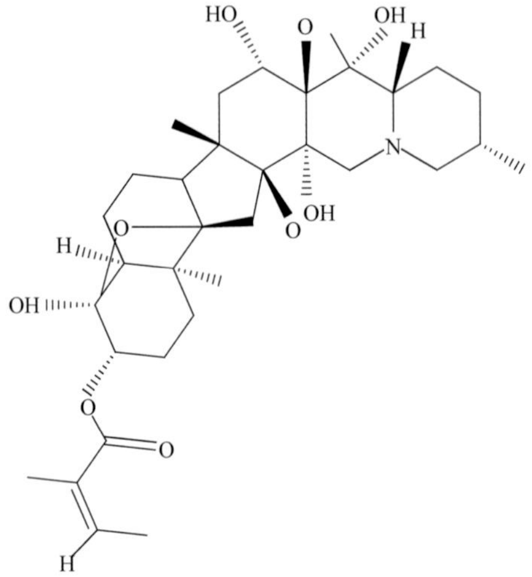

Fig. 321. Structure of cevadine.

Further Reading

Babulova A, Buran L, Selecky FV, "The cardiotoxic activity of helveticoside, a cardiac glycoside from *Erysimum canescens* Roth", *Arzneimittel-Forschung.* 1963; **13**: 412–414.

Bauer S, Bauerova O, Masler L, Sikl D. Erycanoside, a new cardiac glycoside from *Erysimum canescens* Roth. *Experientia.* 1962; **15**: 441.

Feofilatov VV, Loshkarev PM. Erysimine, cardiac glycoside from *Erysimum canescens* Roth. *Dokl Akad Nauk SSSR.* 1954; **94**: 709–12.

Jangwana JS, Bahugunaa RP. Clemontanoside B, A New Saponin from *Clematis montana. Int J Crude Drug Res.* 1990; **28**: 39–42.

Kay JM, Heath D, Smith P, Bras G, Summerell J. Fulvine and the pulmonary circulation. *Thorax.* 1971; **26**: 249–61.

McLean E, Bras G, György P. Veno-occlusive Lesions in Livers of Rats Fed *Crotalaria fulva. Br J Exp Pathol.* 1964; **45**: 242–247.

Schoental R. Alkaloidal Constituents of *Crotalaria fulva* Roxb., Fulvine and Its N-Oxide. *Aust J Chem.* 1963; **16**: 233–238.

Thapliyal RP, Bahuguna RP. An oleanolic acid based bisglycoside from *Clematis montana* roots. *Phytochemistry.* 1993; **34**: 861–862.

Wagenvoort CA, Wagenvoort N, Dijk HJ. Effect of fulvine on pulmonary arteries and veins of the rat. *Thorax.* 1974; **29**: 522–9.

Wang J, Xu QL, Zheng MF, Ren H, Lei T, Wu P, Zhou ZY, Wei XY, Tan JW. Bioactive 30-noroleanane triterpenes from the pericarps of *Akebia trifoliata*. *Molecules*. 2014; **19**: 4301–12.

Wang J, Ren H, Xu QL, Zhou ZY, Wu P, Wei XY, Cao Y, Chen XX, Tan JW. Antibacterial oleanane-type triterpenoids from pericarps of *Akebia trifoliata*. *Food Chem*. 2015; **168**: 623–9.

Index